Products from Alkanes, Cellulose and other Feedstocks

Products from Alkanes, Cellulose and other Feedstocks

Managing Editor: A. Fiechter

with 57 Figures

Akademie-Verlag · Berlin 1981

Die Originalausgabe erscheint im Springer-Verlag
Berlin—Heidelberg—New York
Vertrieb für alle Staaten mit Ausnahme der sozialistischen Länder: Springer-Verlag
Berlin—Heidelberg—New York
Vertrieb für die sozialistischen Länder: Akademie-Verlag · Berlin

Im Springer-Verlag erschienen in der Schriftenreihe
Advances in Biochemical Engineering, Volume 17

Erschienen im Akademie-Verlag, DDR-1080 Berlin, Leipziger Straße 3—4

Lizenznummer: 202 · 100/518/81
Gesamtherstellung: VEB Druckerei „Thomas Müntzer", ·
DDR-5820 Bad Langensalza
Bestellnummer: 762 941 2 (6620) · LSV 1315
Printed in GDR
DDR 76,— M

Contents

Production of Useful Compounds from Alkane Media in Japan 1
S. Fukui, A. Tanaka, Kyoto (Japan)

Microbial Reactions in Prostaglandin Chemistry 37
J. Jiu, Chicago (USA)

Methanol as Carbon Source for Biomass Production in a Loop Reactor 63
U. Faust, W. Sittig, Frankfurt/M. (Germany)

Properties and Mode of Action of Cellulase 101
Y.-H. Lee, L. T. Fan, Manhattan, Kansas (USA)

Kinetics of Hydrolysis of Insoluble Cellulose by Cellulase 131
Y.-H. Lee, L. T. Fan, L. S. Fan, Manhattan, Kansas (USA)

Author-Index 1–17

Production of Useful Compounds from Alkane Media in Japan

S. Fukui and A. Tanaka
Laboratory of Industrial Biochemistry, Department of Industrial Chemistry, Faculty of Engineering, Kyoto University, Yoshida, Sakyo-ku, Kyoto 606, Japan

1 Introduction ... 2
2 Amino Acids ... 5
2.1 L-Glutamic Acid ... 6
2.2 L-Lysine ... 7
2.3 L-Threonine ... 8
2.4 L-Isoleucine ... 8
2.5 L-Valine ... 8
2.6 L-Serine ... 8
2.7 L-Homoserine ... 9
2.8 L-Ornithine ... 9
2.9 L-Citrulline ... 9
2.10 L-Thyrosine ... 9
2.11 L-Phenylalanine ... 9
2.12 L-Alanine ... 9
2.13 Other Amino Acids ... 10
3 Organic Acids ... 10
3.1 α-Ketoglutarate ... 10
3.2 Citrate and Isocitrate ... 10
3.3 C_7-Acids ... 12
3.4 Fumarate, Malate and Succinate ... 12
3.5 Anglyceric Acid ... 12
3.6 Long-chain Dicarboxylic Acids ... 13
4 Carbohydrates and Related Compounds ... 14
4.1 Glucose, Trehalose, Trehalose Lipid and Rhamnolipid ... 15
4.2 Polysaccharides ... 15
4.3 Polyols ... 15
5 Nucleic Acids and Related Compounds ... 17
6 Vitamins, Coenzymes and Related Compounds ... 17
6.1 Vitamin B_2 ... 18
6.2 Vitamin B_6 ... 18
6.3 Vitamin B_{12} ... 18
6.4 Biotin-Vitamer ... 19
6.5 Coenzyme A ... 19
6.6 Cytochrome c ... 19
6.7 Porphyrins ... 19
6.8 Pteridines ... 20
6.9 Ergosterol ... 20
6.10 Carotenoids and Xanthophylls ... 20
6.11 Coenzyme Q ... 21
7 Antibiotics ... 22
7.1 Phenazine Derivatives ... 22
7.2 β-Lactam Antibiotics ... 22
7.3 Peptide Antibiotics ... 23
7.4 Corynecins ... 23
7.5 Miscellaneous ... 25

8 Enzymes and Proteins 25
8.1 Protease 25
8.2 Lipase 26
8.3 Peroxisomal Enzymes 26
8.4 Protein-like Activator 28
9 Biomass with Low Content of Odd-chain Fatty Acids 29
10 References 30

A variety of useful compounds, such as amino acids, organic acids, carbohydrates, nucleotides, vitamins and coenzymes, antibiotics and biomass *etc.*, can be produced by microbial processes utilizing alkanes as substrates.

These products are classified into three groups. The Group 1 involves compounds which can be easily produced by conventional processes utilizing carbohydrates. The Group 2 contains products, whose formation is significantly favoured by physiological and metabolic features of alkane-utilizers. The Group 3 consists of products, which have structures closely related to alkane substrates and can be specifically produced in alkane media. At present these compounds are rarely produced from alkanes on industrial scales. However, production of compounds belonging to Group 2 and 3 is not only of academic interest but also of commercial importance even if the price of alkanes will be further elevated.

1 Introduction

Extensive studies have been carried out in an attempt to utilize alkanes as carbon sources for the production of a variety of useful compounds, such as cell mass, proteins, carbohydrates, nucleic acids, lipids, amino acids, organic acids, vitamins and coenzymes, antibiotics and so on[1,2]. These products can be classified into three groups from biochemical standpoints.

The first group (Group 1) involves products common to those of conventional microbial processes using carbohydrate substrates. Cell mass, carbohydrates, amino acids, nucleic acids, antibiotics, several kinds of enzymes, organic acids and vitamins listed in Table 1 can be produced from either alkanes or carbohydrates, although the flow of carbon in the metabolism of these substrates is different in some part and similar in other part (Fig. 1), and strains of microorganisms as well as the cultivation procedures are different. This implies that selection of carbon sources for production of the first group compounds would be made according to their cost performances. However, the production of cell mass or proteins from alkanes, for example, has special importance to relieve food shortage from global viewpoints. Amino acids, such as lysine and glutamic acid, and organic acids, such as citric acid, are also promising products in using alkanes.

The compounds belonging to the second group (Group 2) are also common to carbohydrate processes, but their productivities are enhanced markedly by special features in alkane assimilation (Table 2). Effluent formation of acetyl-CoA by oxidative degradation of alkanes facilitates biosyntheses of isoprenoids, such as carotenoids, xanthophylls, steroids and coenzyme Q. Alkane substrates provide hydrophobic environment which would favour the production of such water-insoluble compounds. Oxidative degradation of alkanes through β-oxidation enhances

Table 1. Products common to carbohydrate substrates (Group 1)

L-*Amino Acids*: Glutamate, Alanine, Lysine, N^{ε}-Acetyl-lysine, Threonine, Valine, Homoserine, Serine, Phenylalanine, Tyrosine, Ornithine, Citrulline, *etc.*
Organic Acids: α-Ketoglutarate, Citrate, Isocitrate, Fumarate, Malate, Succinate, Anglycerate, *etc.*
Carbohydrates: Trehalose, Arabitol, Erythritol, Mannitol, Polysaccharides, *etc.*
Nucleic Acids, Nucleotides, Nucleosides: AMP, GMP, IMP, cAMP, Inosine, Orotic acid, Orotidine, DNA, RNA, *etc.*
Vitamins and Coenzymes: Vitamin B_6 and Pyridoxal 5′-phosphate, Vitamin B_{12}, Porphyrins, *etc.*
Antibiotics: Pyocyanine, Phenazine-1-carboxylic acid, Oxychlororaphine, Cepharosporins, Corynecins, Brevimycin, Cryomycin, Fluopsins, *etc.*
Enzymes: Lipase, Protease, D-Amino acid oxidase, *etc.*
Cell Mass

the cellular level of CoA as well as those of the enzymes and coenzymes relating to the pathway, such as cytochromes and coenzyme Q (Fig. 2). For example, in alkane media, but not in glucose media the production of coenzyme Q by some yeasts can be enhanced markedly by the addition of *p*-hydroxybenzoic acid, a precursor of the quinone moiety of the coenzyme. Much of the coenzyme formed under these conditions was detected in extramitochondrial portion of the cells. The synthesis of a biotin-vitamer (desthiobiotin) by a strain of *Pseudomonas* is stimulated by odd-chain alkanes, from which pimelic acid, a precursor of the biotin-vitamer, is derived *via* diterminal oxidation followed by subsequent β-oxidation. The most important and interesting characteristics of alkane-utilizing yeasts is the conspicuous appearance of specific cytoplasmic organelles — peroxisomes or microbodies. These peroxisomes contain various specific enzymes, especially those participating in degradation and assimilation of alkanes and those producing hydrogen peroxide. The production of several peroxisomal enzymes will be promising.

The third class (Group 3) includes products, which have structures related closely to alkane substrates, and hence can be specifically produced in alkane media (Table 3). Dicarboxylic acids of various chain length are synthesized directly from alkanes through diterminal oxidation. Mutant strains of yeasts, defective in alkane assimilation or dicarboxylic acid degradation, are used for this purpose. Several peroxisomal enzymes specifically induced by alkanes are also included in this group.

In addition, so-called "co-oxidation" techniques are widely used in the conversion of aliphatic, alicyclic and aromatic hydrocarbons[1,3]. This includes the supply of energy *via* the metabolism of co-substrate or the induction of the substrate-oxidizing

Table 2. Products related to physiological and metabolic features of alkane assimilation (Group 2)

Products Related to Physiological Features
Trehalose lipids, Rhamnolipids, Lipopolysaccharides, *etc.*
Riboflavin, FMN, FAD, Cytochrome c, Coenzyme A, Coenzyme Q, Erogosterol, Carotenoids, *etc.*
Catalase, Uricase, Glycerophosphate dehydrogenase, Lipoproteins, *etc.*
Products Related to Metabolic Features
Biotin-vitamer, Coenzyme Q, Ergosterol, Carotenoids, *etc.*

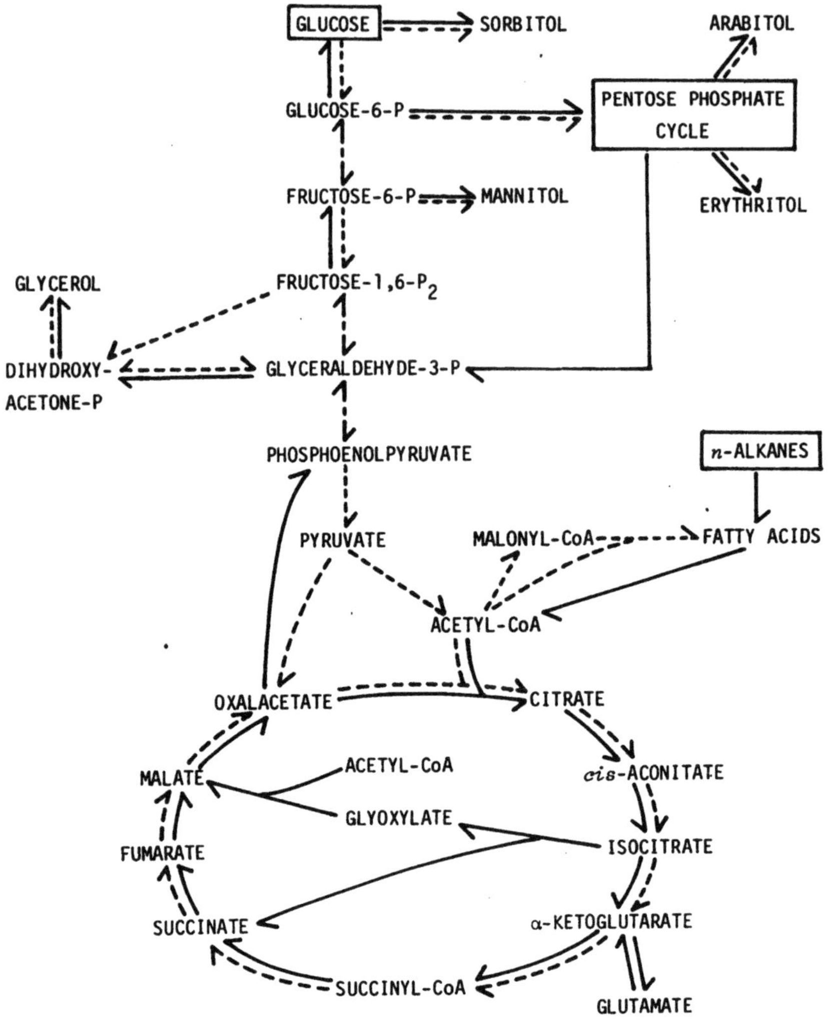

Fig. 1 Metabolic pathways of alkanes and glucose. ————: alkanes; – – – –: glucose

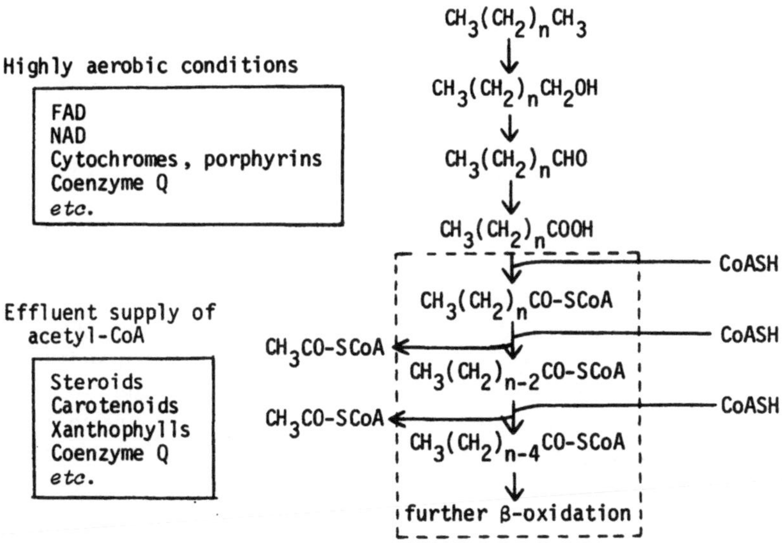

Fig. 2 Relationship of physiological and metabolic features of alkane utilization to microbial products

Table 3. Products specific to alkane substrates (Group 3)

Methylcitrate, Methylisocitrate, *etc.*
Monocarboxylic acids, Dicarboxylic acids, ω-Hydroxy fatty acids, Fatty alcohols, Wax esters, Monoalkenes, *etc.*
Enzymes participating in oxidation and degradation of alkanes

enzymes by co-substrates. A part of the co-oxidation processes will be included in the third group.

This review summarizes the production of useful compounds from alkane substrates by the microbial processes mostly carried out in Japan.

2 Amino Acids

Pioneering work on the amino acid production from alkanes was carried out by Yamada and his coworkers[4,5,6]. They reported that bacterial isolates belonging to the various genera could accumulate more or less a considerable variety of amino acids in kerosene media. Thereafter many research groups have improved the techniques by using special cultivation methods and mutant strains. The productivities of amino acids from alkanes are not necessarily superior to those from carbohydrates and petrochemicals.

2.1 L-Glutamic Acid

Yamada et al.[6] described that Strain S10B1 of *Corynebacterium hydrocarboclastus* produced 485.5 mg l^{-1} of glutamate in a hydrocarbon medium. This bacterium did not require biotin but did require thiamine, unlike glutamate producers from carbohydrates. At the suboptimal concentration of thiamine (5 μg l^{-1}) the productivity of glutamate was enhanced up to 5 g l^{-1} [7]. They further improved the cultural conditions of the bacterium to produce about 6.3 g l^{-1} of the amino acid[8]. The glutamate formation was accompanied by the accumulation of a large amount of α-ketoglutarate[9]. It was supposed that thiamine deficiency interfered with the tricarboxylic acid cycle at the step of α-ketoglutarate dehydrogenase, leading to the accumulation of α-ketoglutarate and subsequently of glutamate[10].

It is well established that the production of glutamate from carbohydrates is successfully achieved under biotin-deficient conditions by using naturally-occurring biotin auxotrophs. In the case of biotin-containing media, the use of oleate auxotrophs or the addition of penicillin gives high yields. These results are considered to be due to the increased excretion of glutamate by the disturbance of the cell membrane synthesis; the oleate auxotrophs lack the ability to synthesize unsaturated fatty acids and penicillin inhibits the complexing of the cell membrane. The effect of penicillin on the glutamate production from alkanes was also confirmed. *Corynebact. hydrocarboclastus* M-104 [11] accumulated 4.0 g l^{-1} of the amino acid with penicillin, the amount being more than ten-fold higher than that without the antibiotic[12]. Hence, the addition of penicillin has been employed as an excellent technique to produce glutamate from alkanes[13,14,15,16,17]. Accumulation of α-ketoglutarate was found to be reduced by the penicillin addition[16]. Under the improved conditions 82 g l^{-1} [17] or 2.3 g l^{-1} h^{-1} of glutamate[18] was produced by *Arthrobacter paraffineus* KY 4303. Decreases of oxygen absorption and substrate consumption which resulted in the decreased glutamate accumulation were often observed in the presence of penicillin. To avoid this phenomenon, penicillin-resistant mutants of *Corynebact. hydrocarboclastus* were isolated[19]. One of the mutants produced 84 g l^{-1} of the amino acid.

Glutamate-producing bacteria isolated hitherto did not show a biotin requirement in alkane media. In the case of alkane-utilizing bacteria, unsaturated fatty acids are directly derived from alkane substrates. These facts indicate that a novel scheme, substituting for the control of the biotin level or for the use of oleate auxotrophs employed in the carbohydrate processes, might be employed for the production of glutamate from alkanes. As mentioned above, addition of penicillin gave good results on the glutamate production. The effect of penicillin on the glutamate excretion was later found to be correlated to the extracellular accumulation of phospholipids[20,21].

Thus, it would be reasonable to predict that reduction in cellular phospholipids, main components of cell membrane, can be achieved by inhibition of glycerol synthesis, bringing about exocellular glutamate accumulation. A glycerol auxotroph (Strain GL-21) was isolated from *Corynebact. alkanolyticum* Strain No. 314 by Nakao et al.[22,23]. The mutant produced about 40 g l^{-1} of glutamate in the presence of a suboptimal level of glycerol (0.1 mg ml^{-1}) and in the absence of penicillin, while the addition of penicillin was necessary for the glutamate production by the strain 314 irrespective of the glycerol concentration or by the strain GL-21 in media containing over 0.3 mg ml^{-1} of glycerol. The strain GL-21 was found to be defective in L-glycerol-3-phosphate:NADP oxidoreductase which is indispensable for the glycerol synthesis[24]. From such results, the permeability to glutamate in the mutant was proved to be enhanced, but the activities of the enzymes participating in glutamate synthesis were comparable to those of the parent strain[25]. Under the optimal conditions the strain GL-21 produced 72 g l^{-1} of glutamate[26]. Kikuchi[27] summarized the results on the glutamate permeation under different conditions employed hitherto (Fig. 3). The method to use the glycerol auxotroph is unique in the utilization of alkanes, although the mutant is also useful for glutamate production from different carbon sources.

Typical data on glutamate production from alkanes are summarized in Table 4.

2.2 L-Lysine

Although accumulation of lysine by bacteria in an alkane medium was first demonstrated by Yamada et al.[5,6], the yield was extremely low (24 mg l^{-1}). Ishii et al.[28] tried to produce various amino acids from alkanes by using auxotrophic mutants of *Corynebact. hydrocarboclastus* R-7 and *Alcaligenes marshallii* P-9. Of the auxotrophs obtained, Strain PN-46 (methionine$^-$) of *A. marshallii* accumulated 0.25 g l^{-1} of lysine. High yields of lysine were achieved by a research group of Kyowa Fermen-

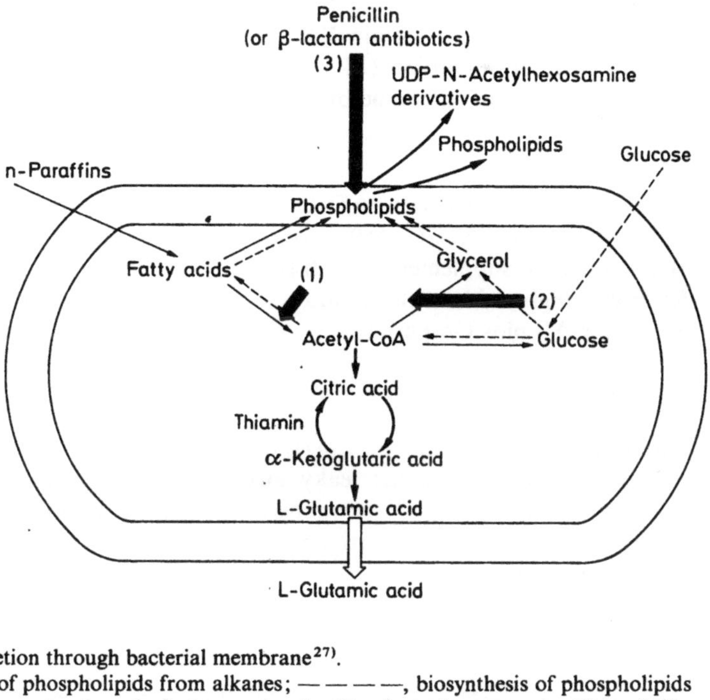

Fig. 3 L-Glutamate excretion through bacterial membrane[27].
————, biosynthesis of phospholipids from alkanes; — — — —, biosynthesis of phospholipids from glucose. (1), biotin auxotroph and oleate auxotroph; (2), glycerol auxotroph; (3) penicillin addition

Table 4. L-Glutamate production from alkanes

Microorganisms	Yield	Ref.
Corynebact. hydrocarboclastus (Thiamine deficiency)	$6 g l^{-1}$	8)
Arthrobacter paraffineus (Penicillin addition)	$82 g l^{-1}$	17)
Corynebact. hydrocarboclastus (Penicillin^r mutant)	$84 g l^{-1}$	19)
Corynebact. alkanolyticum (Glycerol$^-$ mutant)	$72 g l^{-1}$	26)

tation Industry [see[1,29]]. A homoserine or methionine auxotroph of *Brevibacterium ketoglutamicum* produced $10 g l^{-1}$ of lysine, and a homoserine auxotroph of *Nocardia sp.* No. 258 accumulated $34 g l^{-1}$ of the amino acid.

2.3 L-Threonine

An isoleucine leaky auxotroph (Strain KY 7104) of *Arthrobacter paraffineus* accumulated threonine ($9 g l^{-1}$), valine ($9 g l^{-1}$), serine ($2 g l^{-1}$) and leucine ($2 g l^{-1}$) at a relatively high concentration of thiamine[30]. To increase the yield of threonine and to reduce valine accumulation, an isoleucine plus methionine auxotroph (Strain KY 7137) was derived from Strain KY 7104. A revertant with respect to isoleucine requirement (Strain KY 7135), obtained from Strain KY 7137, showed a high yield of threonine ($14.5 g l^{-1}$) and a reduced accumulation of valine[31]. The productivity of threonine by the mutant increased up to $18.9 g l^{-1}$ [see, [1]].

2.4 L-Isoleucine

Micrococcus paraffinolyticus ATCC 15582, *M. paraffineus* KY 4307 and *Corynebact. hydrocarboclastus* ATCC 15592 accumulated $1.55–2.15 g l^{-1}$ of isoleucine in an alkane medium supplemented with DL-α-aminobutyric acid (α-AB) and fumarate[32]. Relatively high yields of the amino acid were also obtained by adding L-homoserine, α-AB or α-AB plus L-aspartate.

2.5 L-Valine

An isoleucine auxotroph of *Corynebact. hydrocarboclastus* accumulated $2.2 g l^{-1}$ of valine[28]. The isoleucine leaky auxotroph of *Arthrobacter paraffineus* (Strain KY 7104) produced $9 g l^{-1}$ of valine accampanied by the production of threonine, as described above[30]. At low concentrations of thiamine the product was converted to glutamate.

2.6 L-Serine

Arthrobacter paraffineus KY 7137 (methionine$^-$, isoleucine$^-$) mentioned above accumulated $1.3 g l^{-1}$ of serine when $10 mg l^{-1}$ of isoleucine and $200 mg l^{-1}$ of methionine were supplemented to an alkane medium. Addition of glycine (5 g l^{-1}) as a precursor enhanced the serine production up to 3.2 g l^{-1}. An α,ε-diaminopimelic acid plus isoleucine auxotroph of *Arthrobacter paraffineus* was also found to be a serine producer. The accumulation of serine by these mutants was always accompanied by accumulation of threonine[33)].

2.7 L-Homoserine

A threonine auxotroph of *Corynebacterium sp.* KY 7142 produced 12 g l^{-1} of homoserine in an alkane medium supplemented with 1 g l^{-1} of threonine. Smaller amounts of isoleucine and valine were also accumulated together with homoserine[34)].

2.8 L-Ornithine

An arginine auxotroph (Strain RN-362) of *Corynebact. hydrocarboclastus* R-7 accumulated 3.9 g l^{-1} of ornithine in an alkane medium[28)]. By the optimization of the cultural conditions the yield was increased up to 9.3 g l^{-1} [35)]. An arginine auxotroph of *Arthrobacter paraffineus* was also utilized for production of ornithine[36)]. The maximal yield of ornithine was 8.0 g l^{-1}.

2.9 L-Citrulline

An arginine requiring mutant of *Corynebacterium sp.* KY 4442 accumulated 8 g l^{-1} of citrulline in an alkane medium supplemented with 1 g l^{-1} of arginine[37)]. Addition of several amino acids or casamino acid was effective for the production of citrulline. An arginine auxotroph of *Arthrobacter paraffineus* produced 10.2 g l^{-1} of citrulline [see, [1)]].

2.10 L-Tyrosine

A phenylalanine auxotroph of *Corynebact. hydrocarboclastus* R-7 was tested for the production of tyrosine from alkanes, but the productivity was rather low (1.0 g l^{-1}) [28)]. A phenylalanine auxotroph of *Corynebacterium sp.* KY 4336 accumulated about 19 g l^{-1} of the amino acid when the pH of the medium was shifted from 6.8 to 7.5 at the exponential growth phase[38)]. This effect of the pH shift might reflect the activity of transamination of *p*-hydroxyphenylpyruvate to tyrosine, since the accumulation of the acid was high at the acidic pH.

2.11 L-Phenylalanine

A tyrosine auxotroph of *Corynebacterium sp.* KY 4309 produced 10 g l^{-1} of phenylalanine at pH 6.0[39)]. When the pH of the medium was kept at 7.0, however, the yield of phenylalanine decreased, and there was accumulation of phenylpyruvate.

2.12 L-Alanine

An arginine auxotroph of *Corynebact. hydrocarboclastus* R-7 accumulated 0.8 g l^{-1} of alanine[28)]. A glutamate-producer, *Corynebact. hydrocarboclastus* S10B1 formed 1.5 g l^{-1} of alanine under the thiamine deficient conditions with concomitant accumulation of large amounts of glutamate and α-ketoglutarate[9]. A strain of *Corynebact. hydrocarboclastus* also produced 4 g l^{-1} of alanine [see, [29]].

2.13 Other Amino Acids

Productions of several amino acids, such as L-tryptophan, L-leucine and L-proline were also examined. Anthranilic acid was found to be effective for the accumulation of tryptophan in alkane media. *Brevibact. ketoglutamicum* (1.5 g l^{-1}) and *Candida tropicalis* (2.0 g l^{-1}) were good producers of tryptophan [see, [1]]. An isoleucine auxotroph of *Corynebact. hydrocarboclastus* R-7 produced 0.7 g l^{-1} of leucine[28]. An arginine auxotroph of the same bacterium accumulated 0.3 g l^{-1} of proline[28].

A large amount of ε-N-acetyl-L-lysine was produced by a lysine-producing mutant (Strain CR-27) of *Corynebact. hydrocarboclastus* KY 8837. The mutant showed a decreased ability to utilize acetyl-lysine as carbon source. Under the optimal conditions Strain CR-27 accumulated 41 g l^{-1} of acetyl-lysine and 9 g l^{-1} of lysine[40].

3 Organic Acids

Accumulation of oxidation products of alkanes has been reported since the early stage of petroleum microbiology [see, [1]]. However, most of these products were not regarded to be useful because of their low yields. In the course of studies on glutamate production from alkanes, concomitant accumulation of α-ketoglutarate was discovered. This finding emphasized possibilities of production of various acids metabolically related to α-ketoglutarate from alkanes. Several acids belonging to the tricarboxylic acid cycle, especially citrate, were found to be efficiently accumulated in alkane media. In addition to these organic acids of the Group 1 (*loc. cit.*), production of long-chain dicarboxylic acids was attempted. These acids belonging to the Group 3, compounds derived directly from alkanes, are promising products.

3.1 α-Ketoglutarate

A glutamate-producing bacterium *Corynebact. hydrocarboclastus* S10B1 accumulated 16 g l^{-1} of α-ketoglutarate under thiamine deficient conditions[9]. Similarly, *Corynebacterium sp.* KY 4439, a glutamate-producer, accumulated the acid with the yield of 85.8% from 8.7% of alkanes[41] at suboptimal concentrations of thiamine. Various strains of *Candida lipolytica* also accumulated α-ketoglutarate in an alkane medium. Among them Strain AJ 5004 produced 65.8 g l^{-1} of the acid[42,43,44]. In these processes the level of thiamine in the medium was kept low.

3.2 Citrate and Isocitrate

Citrate is a promising product from alkanes, because the acid is widely utilized in the food, pharmaceutical and chemical industries.

Tabuchi et al.[45] reported that *C. lipolytica* Strain No. 228, which had been isolated as a potent citrate-producer from glucose, converted 56% of the substrate *n*-hexadecane to yield 34 g l^{-1} of citrate. They also found that various strains of *Candida* yeasts have the ability to accumulate isocitrate in addition to citrate in an alkane medium[46]. The ratio of citrate to isocitrate varied depending on cultural conditions. Reduction of iron concentration in a medium resulted in an increase in the citrate production and in the decrease in the isocitrate accumulation. Under these conditions *C. lipolytica* No. 6–20 produced 85–90 g l^{-1} of citrate and about 20 g l^{-1} of isocitrate from 60 g l^{-1} of alkane mixture, whereas an iron sufficient medium gave about 50 g l^{-1} of citrate and 40 g l^{-1} of isocitrate[47]. When $CaCO_3$ was omitted from the medium and the N/C ratio was increased, the productivity of citrate decreased, while the cell yield increased significantly. Thus, it is necessary for obtaining citrate in a high yield to keep the medium at near neutral pH and at a low nitrogen concentration[48]. The effect of ferrous ion deficiency on the citrate accumulation was ascribed to a reduced activity of aconitase, an iron-containing enzyme. Thiamine deficiency was reported to enhance the isocitrate dehydrogenase activity of the yeasts, resulting in the accumulation of α-ketoglutarate[49]. A problem in the citrate production by yeasts is the concomitant accumulation of isocitrate as described above, which reduces the yield of citrate and makes the purification of citrate difficult. *C. lipolytica* showed a relatively high level of aconitase, which could be inhibited by monofluoroacetate (MFA). In fact, the addition of MFA to the medium decreased the accumulation of isocitrate markedly[50].

Akiyama et al.[51,52,53] isolated two mutant strains from citrate-producing *C. lipolytica* IFO 1437. These mutants assimilated citrate only slightly as a sole carbon source, and one of them, Strain S-22, was very sensitive to MFA. Aconitase activity of another strain, K-20, was one-tenth of that of the parent strain, while the activity of the strain S-22 was about one-hundredth. In accordance with the decreased aconitase activity the productivity of citrate increased from 60 g l^{-1} with the parent strain to 110 g l^{-1} with the strain S-22. The ratio of citrate to isocitrate decreased markedly, from 60:40 with the parent to 97:3 with the mutant. The results reported by Akiyama et al. summarized in Fig. 4 together with those of Tabuchi et al.

With *C. zeylanoides* the sum of citrate and isocitrate, produced from 7.75% of *n*-alkanes, reached 150%, but the ratio of these acids was about 50:50[54]. The pH control with ammonia water gave a good result on the citrate production, 99 g l^{-1} of citrate and 11 g l^{-1} of isocitrate[55].

One mutant of *C. lipolytica* produced 170 g l^{-1} of citrate and 4 g l^{-1} of isocitrate[56]. The same research group reported the maximum citrate yield, 183 g l^{-1}

$$\text{cell mass} \xleftrightarrow{\text{high (N/C ratio) low}} \begin{matrix}\text{citrate}\\ \text{isocitrate}\end{matrix} \xleftrightarrow{\text{sufficient (thiamine) deficient}} \alpha\text{-ketoglutarate}$$

Parent strain	Citrate, 60 g l^{-1}; isocitrate 40 g l^{-1}
Citrate-nonutilizable mutant	Citrate, 90 g l^{-1}; isocitrate 10 g l^{-1}
Monofluoroacetate-sensitive mutant	Citrate, 110 g l^{-1}; isocitrate 4.5 g l^{-1}

Fig. 4 Citrate production from alkanes by *Candida lipolytica*

(isocitrate, 40 g l^{-1}), with *C. hitachinica* [see, [57]]. A mutant of *C. citrica* also produced citrate in a good yield (102 g l^{-1}) with a reduced accumulation of isocitrate (2 g l^{-1})[58]. Kinetic studies on the citrate and isocitrate production by *C. lipolytica* were carried out'in an NH_4^+-limited chemostat culture[59].

Tabuchi and Hara[60] tried to produce isocitrate from alkanes by using MFA-resistant mutants of *C. lipolytica.* Although the mutants showed slightly higher aconitase activity than the parent strain, the yield of isocitrate was not enhanced. These mutants accumulated an oxyacid which was identified later as 2-methyl-isocitrate.

3.3 C_7-Acids

As mentioned above, MFA-resistant mutants of *C. lipolytica* did not show an increase in the productivity of isocitrate, but accumulated an oxyacid[60], which was identified as 2-methylisocitrate[61,62]. One of the mutants, No. 2 of *C. lipolytica*, accumulated about 35 g l^{-1} of methylisocitrate from odd-chain alkanes, the amount being comparable to the sum of citrate and isocitrate produced from these substrates. Even-chain alkanes and glucose did not serve as substrates for the methylisocitrate production[63]. Furthermore, Tabuchi et al.[64] demonstrated accumulation of methyl-citrate and 2-methyl-*cis*-aconitate by the mutant grown on alkane mixture. The results suggested that propionyl-CoA derived from odd-chain alkanes by β-oxidation might be the precursor of these C_7-acids. In fact, they identified special enzymes, methylcitrate synthase and methylisocitrate lyase, and proposed so-called "methylcitric acid cycle" (Fig. 5) for the metabolism of propionyl-CoA in yeasts[65]. This cycle seems to be specific to the alkane metabolism, and therefore, these C_7-acids belong to the Group 3 compounds.

3.4 Fumarate, Malate and Succinate

Yamada et al.[66] isolated a strain of *C. hydrocarbofumarica* which produced 39 g l^{-1} of fumarate from alkanes. By optimization of the cultural conditions the productivity was enhanced up to 50 g l^{-1} (Y = 84%)[67]. Fumarate thus accumulated was further converted to a more useful organic acid; *l*-malate, by the associated cultivation of *C. hydrocarbofumarica* with *C. utilis* or *Pichia membranaefaciens* having a high fumarase activity[68]. About 30 g l^{-1} of fumarate (Y = 65%) was also produced by *C. blankii* MT 1025 [69].

As described above, yeasts having a high fumarase activity was used for the bioconversion of fumarate produced by *C. hydrocarbofumarica* in an alkane medium. The yield of malate based on alkane substrate was 71–72% (36–38 g l^{-1})[68]. Direct production of malate was also examined. *C. brumptii* IFO 0731 accumulated 24.5 g l^{-1} of malate (Y = 80%)[70].

Succinate was produced by *C. brumptii* IFO 0731 [71,72,73]. The maximal yield was 23.6 g l^{-1} (Y = 67%).

3.5 Anglyceric Acid

C. tenuis IFO 1303 accumulated about 10 g l^{-1} of anglyceric acid in an alkane medium[74]. Relatively high concentrations of $(NH_4)_2SO_4$ was favourable to produce this acid by reducing the accumulation of citrate.

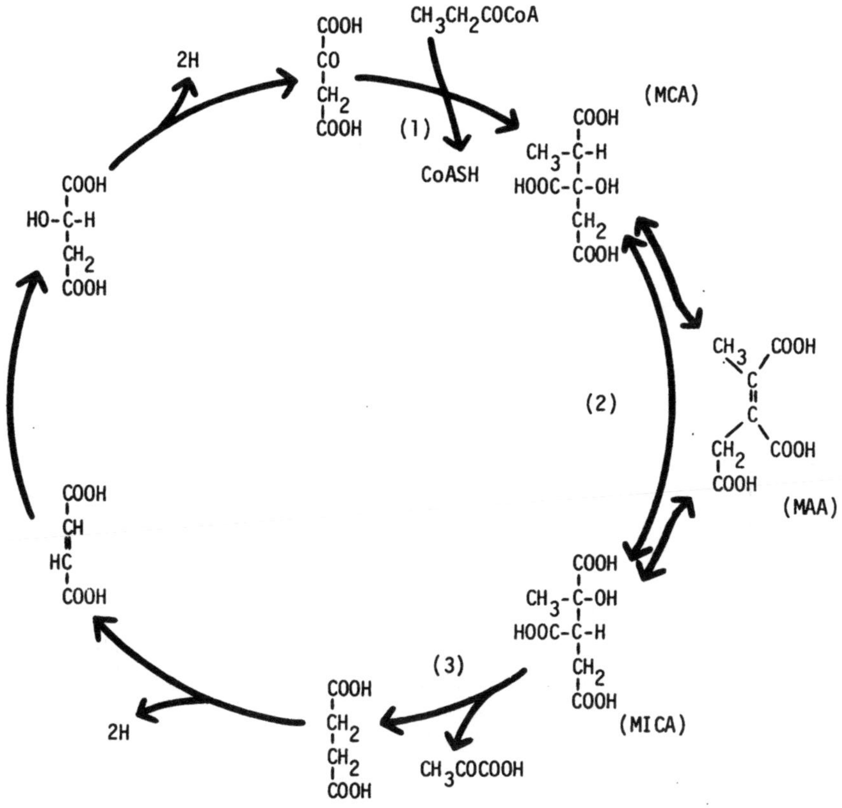

Fig. 5 Methylcitric acid cycle in *Candida lipolytica*[65].
(1), methylcitrate synthase; (2), aconitase; (3), methylisocitrate lyase. MCA, methylcitrate, MAA, methylaconitate; MICA, methylisocitrate

3.6 Long-chain Dicarboxylic Acids

Oxidation products directly derived from alkanes are the typical examples of the Group 3 compounds. Dicarboxylic acids of long-chain, especially tetradecane-1,14-dicarboxylic acid (DC-16), are very valuable raw materials for the preparation of perfumes[75]. Thus, the production of these acids from alkanes by microbial processes has been investigated extensively.

Several strains of yeasts were found to accumulate dicarboxylic acids: *Pichia sp.* Strain Y-3 produced nonane-1,9-dicarboxylic acid (DC-11), azelaic acid (DC-9), pimelic acid (DC-7) and glutaric acid (DC-5) from undecane[76]; *C. rugosa* JF 101 accumulated sebacic acid (DC-10), suberic acid (DC-8), adipic acid (DC-6) and succinic acid (DC-4) from decane[77]; *C. tropicalis* Strain OH 23 produced DC-5 and DC-6 from pentadecane and hexadecane, respectively[78]; *Torulopsis candida* No. 99 accumulated 5.8 g l^{-1} of DC-10 from decane and 0.88 g l^{-1} of DC-9 from nonane. However, alkanes of more than C_{11} gave DC-7 or DC-8 instead of dicarboxylic acids having the same carbon skeleton to the substrates[79].

Shiio and Uchio tried to produce longer-chain dicarboxylic acids, such as DC-16, a precursor of synthetic muscone, from hexadecane. They selected *C. cloacae* 310 as a potent producer of dicarboxylic acids. Although this strain produced shorter-chain dicarboxylic acids (DC-12, 2.28; DC-8, 0.82; DC-6, 1.86 (g l^{-1})) in a dodecane medium, DC-16 was not accumulated in a hexadecane medium (Table 5)[80]. To obtain high yields of dicarboxylic acids and to eliminate the degradation of these acids *via* β-oxidation, a mutant strain M-1 lacking the ability to assimilate dicarboxylic acids was derived from *C. cloacae* 310[81]. The mutant produced enhanced amounts of dicarboxylic acids with the same carbon skeleton of the substrates (nonane to octadecane), and the accumulation of the β-oxidation products was very low. Production of DC-16 from hexadecane reached 29.3 g l^{-1} with the resting cells of the mutant and 21.8 g l^{-1} with the growing cells, acetate being used as co-substrate in both cases to give the reducing equivalent for the hydroxylase reaction. Furthermore, a mutant strain MR-12, which was unable to assimilate alkanes, was derived from *C. cloacae* M-1. The dicarboxylic acid degradation activity of the strain MR-12 was 10% of that of the parent strain, whereas that of the strain M-1 was 40%. The resting cells of the strain MR-12 produced 42.7 g l^{-1} of DC-16 from hexadecane[82] (Table 5). The carbon balance in hexadecane oxidation with the resting cells of the strain MR-12 showed that about 60% of consumed carbon was recovered as DC-16 and about 40% as CO_2. The evolution of CO_2 under these conditions is a very curious phenomenon. After optimization of the cultural conditions the mutant MR-12 accumulated 61.5 g l^{-1} of DC-16 from hexadecane in the medium containing acetate as co-substrate[83]. The scale-up of the process to a 300-l tank has been accomplished[84]. Dicarboxylic acids with chain-lengths of C_{12}–C_{18} were also produced effectively by this mutant.

Table 5. Production of higher dicarboxylic acids from *n*-hexadecane by *Candida cloacae*

Parent strain	
↓ DC-6, 0.21 g l^{-1}; DC-8, 0.07 g l^{-1}; DC-16, 0.004 g l^{-1}	
DC-10-nonutilizable mutant	(acetate as co-substrate)
↓ DC-14, 0.01 g l^{-1}; DC-16, 18.5 g l^{-1}	
Alkane-nonutilizable mutant	(acetate as co-substrate)
↓ DC-16, 42.7 g l^{-1}	
(Optimization of cultural conditions)	
DC-16, 61.5 g l^{-1}	

DC, dicarboxylic acid

4 Carbohydrates and Related Compounds

Conversion of alkanes to carbohydrates, especially polysaccharides, by microorganisms is thought to be one of the most important and interesting processes from both biochemical and practical viewpoints.

4.1 Glucose, Trehalose, Trehalose Lipid and Rhamnolipid

In alkane media, glutamate-producing *Arthrobacter paraffineus* was found to accumulate anthrone-positive substances, which were identified as α,α-trehalose and D-glucose. Addition of penicillin enhanced the yield of trehalose up to 7 g l^{-1}, but did not affect the glucose accumulation (1.8 g l^{-1})[85]. By addition of a trace amount of Cu^{2+}, the maximal yield of trehalose was markedly increased, reaching 11.2 g l^{-1} with 78 g l^{-1} of glutamate[17]. Trehalose lipid composed of trehalose and α-branched-β-hydroxy fatty acid has been often found in the emulsion layer consisting of bacterial cells, lipids and alkanes. The glycolipid was produced not only by *Arthrobacter paraffineus* (1 g l^{-1} as glucose) but also by a variety of alkane-utilizing bacteria[86]. Penicillin addition reduced the accumulation of the glycolipid. Such glycolipids are supposed to serve as surfactants in the bacterial utilization of alkanes, as the case of rhamnolipid produced by *Pseudomonas aeruginosa* in an alkane medium[87]. Rhamnolipids produced from alkanes by *Ps. aeruginosa* were shown to have a bactericidal activity[88]. These sugar lipids are examples of the Group 2 compounds.

4.2 Polysaccharides

Although production of polysaccharides from various petrochemicals has been documented, little information is available with respect to that from alkanes.

Corynebacterium Strain No. 1645 and *Brevibacterium* Strain No. 1625-1 accumulated acid heteropolysaccharides from alkanes. The saccharide produced by the former strain consisted of glucose and mannose as sugar moieties, and lactic, levulinic, succinic and lauric acids as acid moieties. The saccharide from the latter strain was composed of glucose, mannose and galactose as sugar moieties, and lactic, levulinic and lauric acids as acid moieties[89].

4.3 Polyols

Several investigations were carried out concerning the conversion from the citrate to the polyol process with alkane-utilizing yeasts.

Tabuchi and Hara found that *C. lipolytica* accumulated mannitol and a small amount of erythritol in an alkane medium when the neutralization of the culture medium with $CaCO_3$ was omitted. The yield of the polyols was about 25 g l^{-1} [90]. A variety of yeasts could produce polyols from alkanes under such conditions.

Hattori and Suzuki investigated in more detail the production of polyols from alkanes. A glycerol auxotroph (Strain KY 6166) of *C. zeylanoides* was selected as an excellent polyol producer. This strain accumulated 55 g l^{-1} ($Y = 55\%$) of *meso*-erythritol in an alkane medium. It was necessary for the erythritol production to keep the medium pH at a low level (2.5–4.0) and the concentration of NaCl or KCl at a high level (1–3%)[91]. The effect of pH on the production of erythritol and of citrate is depicted in Fig. 6. Under the optimal conditions about 180 g l^{-1} of erythritol was produced from alkanes by this yeast strain[92]. When the concentration of phosphate was kept at a high level (about 0.1% as KH_2PO_4), the product of this yeast converted almost completely from erythritol to mannitol (Fig. 7). The yield of mannitol reached 63 g l^{-1} ($Y = 52\%$)[92].

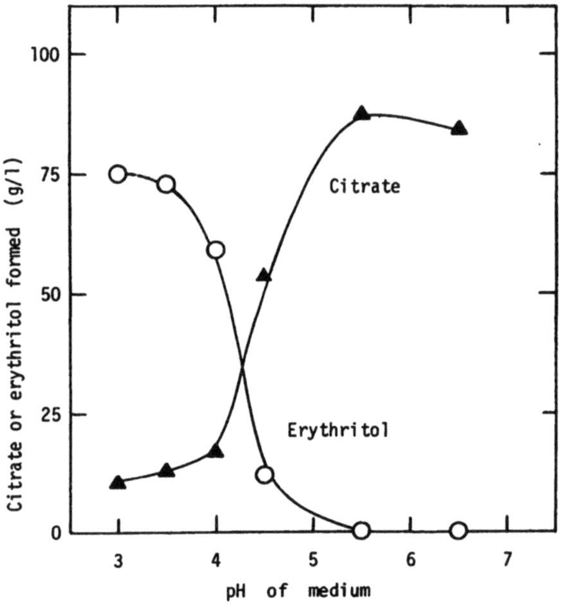

Fig. 6 Effect of medium pH on interconversion between citrate production and erythritol production by *Candida zeylanoides* from alkanes (data obtained from[92])

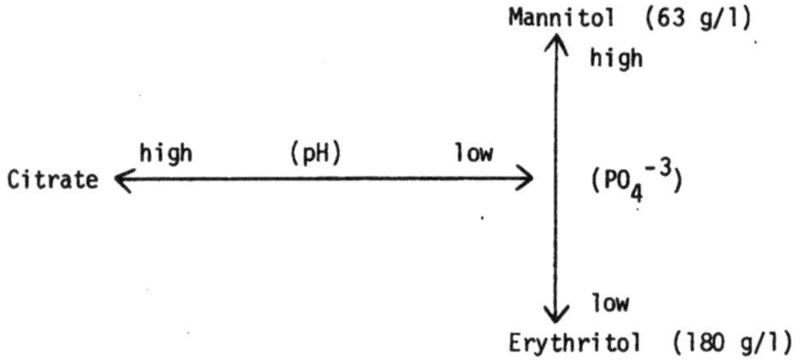

Fig. 7 Effect of cultivation conditions on polyol production from alkanes by *Candida zeylanoides*

The intracellular pH of the polyol-producing cells was found to be about 5.5 (medium pH, below 4.0), while that of the citrate-producing cells was about 6.5 (medium pH, above 4.5). At pH 5.5, the activity of citrate synthase was one-third that at pH 6.5, but the activities of isocitrate lyase and phosphoenolpyruvate carboxykinase were significantly high at pH 5.5. Aconitase showed a similar activity at both pHs. These facts would offer a probable explanation for the mechanism of the conversion from the citrate to the polyol process by lowering the pH of the medium — that is, the reduced accumulation of citrate and the enhanced synthesis of carbohydrates[93]. Interconversion between the erythritol and the mannitol process was ascribable to the controlling effect of phosphate ion on transketolase activity, which would affect the levels of erythrose-4-phosphate and fructose-6-phosphate[94]. On the other hand, *C. tropicalis* KY 6224 accumulated D-arabitol in an alkane medium under the conditions where *C. zeylanoides* KY 6166 produced erythritol (low medium pH and high level of NaCl or KCl). The productivity of arabitol by the yeast was 75 g l^{-1} (Y = 50%)[95].

5 Nucleic Acids and Related Compounds

Inosine is the precursor of 5′-inosinic acid, which is utilized as a seasoning. An adenine auxotroph U-1542 derived from glutamate-producing *Corynebact. petrophilum* SB 4082 accumulated 1.6 g l^{-1} of inosine and 0.3 g l^{-1} of hypoxanthine in an alkane medium[96,97,98]. Glucose did not serve as a substrate for the production of inosine by the mutant[99]. Adenine auxotrophs of *Corynebact. simplex*, *Brevibact. ketoglutamicum* and *Arthrobacter paraffineus* produced 1.5–2.8 g l^{-1} of inosine [see, [1]].

A uracil auxotroph (KY 7122) of *Arthrobacter paraffineus* ATCC 15591 produced 6 g l^{-1} of orotate and 3.5 g l^{-1} of orotidine from alkanes[100]. Similarly, uracil-requiring mutants of *Brevibact. ketoglutamicum* and *Corynebacterium sp.* ATCC 21178 accumulated these compounds [see, [1]].

Nucleosides were prepared from RNA of alkane-grown cells of *Pseudomonas sp.*, which contained 12% of RNA. That is, RNA extracted with 1% NH_3 at room temperature for 24 h was hydrolyzed by 2% NH_3 at 80 °C for 2 h. The 3′-nucleotides thus obtained were converted to nucleosides with a yield of 80–90% by heat treatment at a weak akaline condition (130 °C, pH 7–8)[101].

Salvage syntheses of 5′-nucleotides, such as 5′-inosinic, 5′-guanilic and 5′-adenylic acids, were carried out by using bacteria, yeast and mold. *Arthrobacter simplex* produced 9.9 g l^{-1} of guanilic acid, 1.9 g l^{-1} of inosinic acid and 1.3 g l^{-1} of adenylic acid in the alkane media supplemented with respective bases [see, [1]]. Cyclic adenosine-3′,5′-monophosphate was produced extracellularly in an alkane medium by *Arthrobacter roseoparaffineus* KY 4301 (1.4 g l^{-1}) and by *Micrococcus paraffinolyticus* KY 4306 (0.5 g l^{-1})[102]. *Candida sp.* N-25-2 was shown to have a potent ability to phosphorylate AMP to ATP. Alkane-grown cells converted 20 mM AMP to ATP within 3 h in the presence of phosphoenolpyruvate[103].

Pseudomonas strains were found to accumulate appreciable amounts of DNA (0.1–0.5 g l^{-1}) extracellularly in an alkâne medium[104]. Under optimal conditions 5.5 g l^{-1} of DNA was produced from alkanes. Most of DNA thus excreted was highly polymerized (4×10^6 daltons)[105]. Productivities of RNA by alkane-utilizing *Achromobacter nucleoacidives* No. BA108 were compared in various cultivation methods in a 200 l-reactor. The productivities of RNA in semi-, single-stage and two-stage continuous cultivations were 1.17, 2.99 and 2.24 g l^{-1} h^{-1}, respectively[106].

6 Vitamins, Coenzymes and Related Compounds

As illustrated in Fig. 2, alkanes are usually degraded to produce acetyl-CoAs *via* the β-oxidation pathway under highly aerobic conditions. These conditions would enhance the cellular level of coenzyme A (CoA) and of coenzymes relating to respiratory chain such as cytochromes, coenzyme Q, flavin nucleotides, *etc*. Moreover, an abundant supply of acetyl-CoAs would favour the biosyntheses of isoprenoid compounds — carotenoids, xanthophylls, steroids and the side chain of coenzyme Q, *etc*. Furthermore, alkane substrates give a hydrophobic environment, favouring accumulation of these lipophilic products in the culture broth. Thus, most of vitamins and coenzymes are classified in Group 2 (*loc. cit.*).

6.1 Vitamin B_2

Pichia guilliermondii accumulated in an alkane medium 117 mg l^{-1} of vitamin B_2 mostly composed of riboflavin[107]. The yield of the vitamin was enhanced to 281 mg l^{-1} by the optimization of cultural conditions. When the aeration rate was reduced, excretion of the vitamin into the medium decreased markedly, while the intracellular level of the vitamin increased drastically[108]. *C. tropicalis* pK 233 accumulated about 3 mg l^{-1} of flavin, and a part of the vitamin excreted was found to be the coenzyme form[109]. *Bacterium anitratum* No. S5[110], which was later designated as *Acinetobacter anitratum*[111], accumulated 27 mg l^{-1} of coenzyme forms of flavin, consisting of 47% of FMN and 53% of FAD. This bacterium seems to be unique in its potent extracellular accumulation of coenzyme forms of the vitamin[112].

6.2 Vitamin B_6

C. albicans formed 0.5–0.6 mg l^{-1} of vitamin B_6 in an alkane medium. About 0.2 mg l^{-1} of pyridoxal 5′-phosphate was found in the culture fluid[113, 114]. *C. guilliermondii* produced about 25 mg l^{-1} of vitamin B_6. The phosphorylated compounds, pyridoxamine phosphate and pyridoxal phosphate, were also accumulated in the medium[115].

6.3 Vitamin B_{12}

Vitamin B_{12} is the most complex non-polymer compound in nature as well as the only naturally occurring cobalt-containing compound. This vitamin is produced industrially only by microbial processes, even though its chemical synthesis has been achieved by tremendous efforts of organic chemists.

Several papers have been reported on the production of vitamin B_{12} from alkanes. *Corynebact. simplex* ATCC 6946 produced about 0.6 mg l^{-1} of vitamin B_{12} from alkanes[116]. Most part of the vitamin was retained in the cells in the form of 5,6-dimethylbenzimidazoylcobamide coenzyme (adenosylcobalamin)[117]. The vitamin production by *Corynebact. simplex* was enhanced by addition of appropriate amounts cobaltous sulfate to the medium. Various kinds of bacteria belonging to *Nocardia*, *Mycobacterium*, *Pseudomonas*, *Bacillus* and *Rhodopseudomonas* accumulated 0.03–0.25 mg l^{-1} of the vitamin[116, 118, 119, 120]. The yields of vitamin B_{12} in alkane media were far lower than those reported in methanol media and carbohydrate media.

6.4 Biotin-Vitamer

Pseudomonas sp. Strain 5-2 accumulated in a kerosene medium about 1 mg l^{-1} of a biotin vitamer, which was proved to be desthiobiotin[121]. Supplementation of pimelic acid (DC-7) or azelaic acid (DC-9), the precursor of the vitamin, was effective in the biotin production by the strain. In this connection, undecane and tridecane were the best substrates for the vitamin production. Furthermore, undecane was found to be converted to azelaic acid and pimelic acid *via* diterminal oxidation and subsequent β-oxidation[122]. During the optimization of the cultural conditions, a small amount of adenine was proved to be effective. About 18 mg l^{-1} of the vitamin was produced in the kerosene medium supplemented with 0.6 g l^{-1} of pimelic acid and 1 g l^{-1} of adenine[123]. This process is one of interesting examples of the direct supply from the substrate alkanes of the precursor of useful products with complex structures.

dl-cis-Tetrahydro-2-oxo-4-*n*-pentyl-thieno-(3,4-d)-imidazoline (*dl*-TOPTI) was converted to *dl*-biotinol and *dl*-biotin in a yield of 60% by alkane-utilizing *Corynebacterium sp.* B-321. Strain M-6318 defective in alkane assimilation was derived from the strain B-321 to enhance the biotin yield. This mutant M-6318 showed a repressed activity of biotin degradation, and converted 80% of *dl*-TOPTI to biotin in an acetate medium supplemented with alkanes[124, 125].

6.5 Coenzyme A

Coenzyme A (CoA) is an important reagent in biological systems. This coenzyme plays an essential role in fatty acid degradation and, therefore, its high productivity has been expected in alkane media.

A variety of microorganisms including bacteria and yeasts produced relatively larger amounts of CoA from alkanes (9–31 mg l^{-1}) than from glucose (3–13 mg l^{-1}). Of the microorganisms tested *Ps. alkanolytica* Pd 192 produced *de novo* 80 mg l^{-1} of CoA from alkanes[126].

6.6 Cytochrome c

Cultivation of microorganisms on alkanes is usually carried out under vigorous aeration. These conditions would favour the production of enzymes and coenzymes related to the respiratory chain.

C. albicans IFO 0583 produced about 1.5 mg (g-cell)$^{-1}$ ot 7.5 mg l^{-1} of cytochrome c from alkanes[127]. The content of cytochrome c in *C. lipolytica* NRRL Y-6795 was rather higher than that of *C. albicans*. *C. lipolytica* produced 12.2 mg l^{-1} or 2.1 mg (g-cell)$^{-1}$ of cytochrome c in an alkane medium[128].

6.7 Porphyrins

Several bacterial strains having the ability to produce vitamin B_{12} accumulated appreciable amounts of porphyrins from alkanes, although the role of porphyrins in alkane assimilation is not clarified yet.

Mycobact. smegmatis accumulated a red pigment in an alkane medium under reduced aeration. This pigment was identified as coproporphyrin III. The amount of the porphyrin accumulated was about 60 μg l^{-1} in the alkane medium and 30 μg l^{-1} in the glucose medium[129, 130].

Corynebact. simplex ATCC 6946 accumulated cobalt-coproporphyrin III in a hexadecane medium supplemented with cobaltous sulfate and an appropriate detergent. The detergent was necessary for excretion of the porphyrin. Reduction of aeration also enhanced the porphyrin accumulation but not the vitamin B_{12} production from alkanes[131].

6.8 Pteridines

A mutant strain No. 4510 requiring hypoxanthine and thiamine was derived from *Achromobacter petrophilum* No. 4017. This mutant accumulated pteridine derivatives in an alkane medium supplemented with guanine or other purine bases. These compounds were identified as 6-(3-indolyl)-8-(1-D-ribityl)-2,4,7-trioxohexahydropteridine and 6-(*p*-oxyphenyl)-8-(1-D-ribityl)-2,4,7-trioxohexahydropteridine. The former compound was produced in an amount of 90 mg l^{-1}, and the latter 54 mg l^{-1} [132, 133, 134, 135]. The function of these compounds in the metabolism of this bacterium seems of great interest, but not evident hitherto.

6.9 Ergosterol

Ergosterol is a physiologically important member of steroid as provitamin D. It would be expected that alkanes might serve as excellent carbon sources for ergosterol production, because sufficient amounts of acetyl-CoAs and lipophilic environment can be provided by using these substrates.

The content of ergosterol in alkane-grown cells of *C. petrophilum* was about 1%, being higher than that in glucose-grown cells[136]. *C. tropicalis* produced about 70 mg l^{-1} of ergosterol, and alkanes were better substrates than glucose[137].

On the other hand, in the case of ergosterol formation by fungi, alkanes were not necessarily superior to glucose as carbon source. *Cladosporium cladosporioides*, *Hormodendrum hordei* and *Penicillium lilacinum* produced from undecane 250 mg l^{-1} (cellular content, 1.9%), 140 mg l^{-1} (2.1%) and 13 mg l^{-1} (1.6%) of ergosterol, respectively. The productivities were rather high, but it was necessary to cultivate the fungi for long period[138, 139]. A kabicidin-resistant mutant of *Fusarium sp.*, which was isolated as a potent producer of alkaline protease, accumulated extracellularly 260 mg l^{-1} of ergosterol in an alkane medium[140].

6.10 Carotenoids and Xanthophylls

Various carotenoids, including β-carotene, are useful as provitamin A and as food colorings. One can expect that their production may be stimulated in alkane media as is case for ergosterol formation, as mentioned above.

Several bacteria, such as *Mycobact. smegmatis*, *Nocardia corallina* and *N. lutea*, were found to accumulate water-insoluble orange-red coloured pigments in an alkane medium. *Mycobact. smegmatis* produced 0.8 mg l^{-1} of the pigments, which were proved to be a mixture of several carotenoids. The main component of the pigments was identified as 4-keto-γ-carotene[141, 142]. Such carotenoid-producing bacteria made up a clump of cells occluding pigments as well as alkanes when grown on alkanes as sole carbon sources. In this case, it would be supposed that a part of the carotenoids produced would be extracted from cells by alkanes, and that carotenogenesis of alkane-utilizing bacteria might be enhanced by being freed from the feed-back control resulting from intracellular accumulation of carotenoids. In fact, carotenoids of alkane-grown *Mycobact. smegmatis* were more readily extracted with organic solvents than those of glucose-grown cells, and the amount of the former was higher than that of the latter[143)].

Brevibacterium KY 4313 produced about 1 mg l^{-1} of carotenoids[144)]. The main component was identified as canthaxanthin (4,4′-diketo-β-carotene)[145)]. The biosynthetic pathway of the carotenoids in this bacterium was also investigated[146)]. *Brevibacterium* Strain No. 103 produced 30 μg $(g\text{-cell})^{-1}$ of pigments, the main components of which were reported to be astacin and astaxanthin[147)].

6.11 Coenzyme Q

Coenzyme Q (CoQ) is one of the most interesting and promising products from alkane media, since this coenzyme is a member of the respiratory chain, and its isoprenoid moiety is derived from acetyl-CoAs which are sufficiently supplied by the degradation of alkane substrates. Now in Japan, CoQ_{10} is widely used as a medicine effective for anemia, hyperpiesia, pyorrhea and dystrophy.

Takeda et al.[148)] first demonstrated production of CoQ from alkanes. That is, *Achromobacter sp.* produced 7.5 mg l^{-1} or 1.5 mg $(g\text{-cell})^{-1}$ of CoQ_9 from hexadecane. The present authors have compared the CoQ productivity of various kinds of yeasts and molds from alkanes. Of the microorganisms tested, *C. tropicalis* pK 233 was selected as a relatively potent producer of CoQ. The CoQ content of this yeast was higher in the alkane-grown cells (about 2 mg $(g\text{-cell})^{-1}$) than in the glucose-grown cells (about 0.6 mg $(g\text{-cell})^{-1}$). The coenzyme isolated from this yeast was identified as CoQ_9[149)]. As mentioned above, during the microbial utilization of alkanes the isoprenoid moiety of CoQ would be synthesized easily from acetyl-CoAs supplied in abundance from alkane substrates, while the synthesis of its quinone moiety might be limited. In fact, addition of *p*-hydroxybenzoic acid (POBA), a close precursor of the quinone moiety, stimulated remarkably the production of CoQ by *C. tropicalis*, the yield reaching 10.7 mg l^{-1} or 2.8 mg $(g\text{-cell})^{-1}$. Such stimulatory effect was also observed at a fairly high concentration of L-tyrosine, but L-phenylalanine was not so effective. It is interesting that POBA showed no effect on CoQ formation by the yeast in a glucose medium[150)]. For the CoQ production alkanes with carbon chain of 10–12 were the best substrates[151)]. Under the optimal conditions *C. tropicalis* accumulated 52 mg l^{-1} of CoQ_9 in the cells by successive additions of POBA. The maximal content of CoQ_9 obtained was 4.7 mg $(g\text{-cell})^{-1}$, which would be the highest value obtained hitherto[152, 153)]. Thus, *C. tropicalis* pK 233 was proved to be the most potent CoQ-producer, but unfortunately the product was CoQ_9, which is not permitted as a medicine in Japan. The CoQ accumulated was localized in the extra-mitochondrial fraction of the alkane-grown *C. tropicalis* cells[154)].

Several strains of *Pseudomonas* produced about 2 mg $(g\text{-cell})^{-1}$ of a mixture of CoQ_8 and CoQ_9 from alkanes. *Candida* yeasts were also demonstrated to accumulate 0.8 mg $(g\text{-cell})^{-1}$ of CoQ_9 [see, [155)]]. These microorganisms accumulated epoxy derivatives of CoQ_8 and CoQ_9, which might be derived from CoQ_8 and CoQ_9 during the cultivation with vigorous aeration[156].

7 Antibiotics

Although different types of antibiotics are produced from alkane media by various microorganisms, the number of the products obtained hitherto was far small compared with those produced from carbohydrate media. Moreover, it is said that antibiotics produced by alkane-utilizing microorganisms are not necessarily specific to alkane substrates.

7.1 Phenazine Derivatives

The production of pyocyanine from alkanes was first demonstrated by Lee and Walden[157]. *Ps. aeruginosa* ATS-14 accumulated 140 mg l^{-1} of this pigment in a kerosene medium under stationary conditions. Alkanes with chain lengths of 15–17 were found to be the best substrates. The accumulation of pyocyanine was fairly low under shaking conditions (2 mg l^{-1}). In the meantime, *Ps. aeruginosa* P-05201 accumulated 128 mg l^{-1} of pyocyanine from alkanes (C_{14}–C_{17}) in shaking cultures[158]. Addition of an appropriate amount of peptone was almost essential for the production of the pigment. Pentadecane served as the best substrate for this strain.

Pseudomonas M3542 accumulated 350 mg l^{-1} of 1-phenazinecarboxylic acid in a kerosene medium[159,160]. *Ps. aeruginosa* HT7B1 accumulated 1600 mg l^{-1} of 1-phenazinecarboxylic acid and 850 mg l^{-1} of oxychlororaphine in an alkane medium of pH 5.5[161]. Glucose did not serve as a good substrate for the pigment production by these two strains.

Arthrobacter paraffineus KY 7134 produced two kinds of pigments from alkanes, especially from octadecane. These pigments were identified as 1,6-dihydroxyphenazine (DHP) and 1,6-dihydroxyphenazine-5,10-di-N-oxide (DHPO). The maximal yields of DHP and DHPO were about 500 mg l^{-1} and 550 mg l^{-1}, respectively. Accumulation of DHPO preceded that of DHP during the cell growth[162].

7.2 β-Lactam Antibiotics

Alkane-utilizing fungi, which were designated as *Paecilomyces carneus* C-2237 and *P. persicinus* C-3009 accumulated cepharosporins as well as penicillin N in an alkane medium. The amount of cepharosporins produced by *P. carneus* from alkanes was comparable to that from glucose, while *P. persicinus* accumulated a significantly less amount of the antibiotics in an alkane medium than in a glucose medium. *P. carneus* produced from alkanes (C_{12}–C_{15}) 120 mg l^{-1} of deacetoxycepharosporin C (DACPC), 25 mg l^{-1} of deacetylcepharosporin C (DCPC) and 110 mg l^{-1} of cepharosporin C (CPC)[163]. Under the optimal conditions about 450 mg l^{-1} of cepharosporins were produced, consisting of CPC, DCPC and DACPC in the ratio of 1:1:2[164].

7.3 Peptide Antibiotics

Streptomyces sp. No. 81[165], which was psychrophilic and was later designated as *Str. griseus* subsp. *psychrophilus* AKU 2881, produced two kinds of peptide antibiotics from alkanes[166]. The one, which was called cryomycin and active against some gram-positive bacteria, was exclusively produced at a cultivation temperature of 0–18 °C[167, 168]. The other, M-81, was produced when the microorganism was cultivated at 20–37 °C, but not below 20 °C. M-81 was also active against some gram-positive bacteria[169]. The yields of cryomycin and M-81 from alkanes were 20 mg l^{-1} and 150 units per ml, respectively[166].

7.4 Corynecins

Suzuki et al. found that *Corynebact. hydrocarbaclastus* produced three kinds of antibacterial compounds which were named corynecins (corynecin I, II and III). Alkanes and sucrose were the best substrates for the corynecin production, and the maximal yield of 3 g l^{-1} of corynecins was obtained in an alkane medium. In addition to these three active corynecins, two kinds of inactive compounds (corynecin IV and V) were also accumulated[170]. These compounds were identified as acyl derivatives of D-(—)-threo-1-*p*-nitrophenyl-2-amino-1,3-propanediol (free base of chloramphenicol) (Fig. 8)[171]. Although the antibacterial activities of corynecins are far weaker than chloramphenicol, these compounds give by hydrolysis nitrophenylserinol, which can be converted to chloramphenicol by chemical synthesis. Thus, the bacterial production of corynecins is thought to open another possibility of chloramphenicol production. A polysaccharide-defective and chloramphenicol-resistant mutant[172] and a polysaccharide-defective, chloramphenicol-resistant and

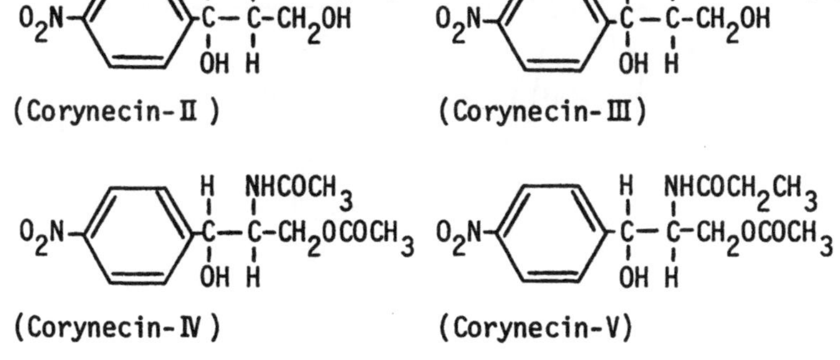

Fig. 8 Structures of chloramphenicol and corynecins

glycolipid-defective mutant were examined for their ability to produce corynecins[173]. Furthermore, a mutant showing an enhanced productivity of corynecin I, a less toxic homolog to the producer, was derived from the latter mutant[173,174]. The production of corynecins was mainly carried out using sucrose[175], but corynecin I was produced successfully from acetate[176].

The biosynthetic pathway of corynecins and its regulation were also investigated [177,178,179]. The results are summarized in Fig. 9[180,181].

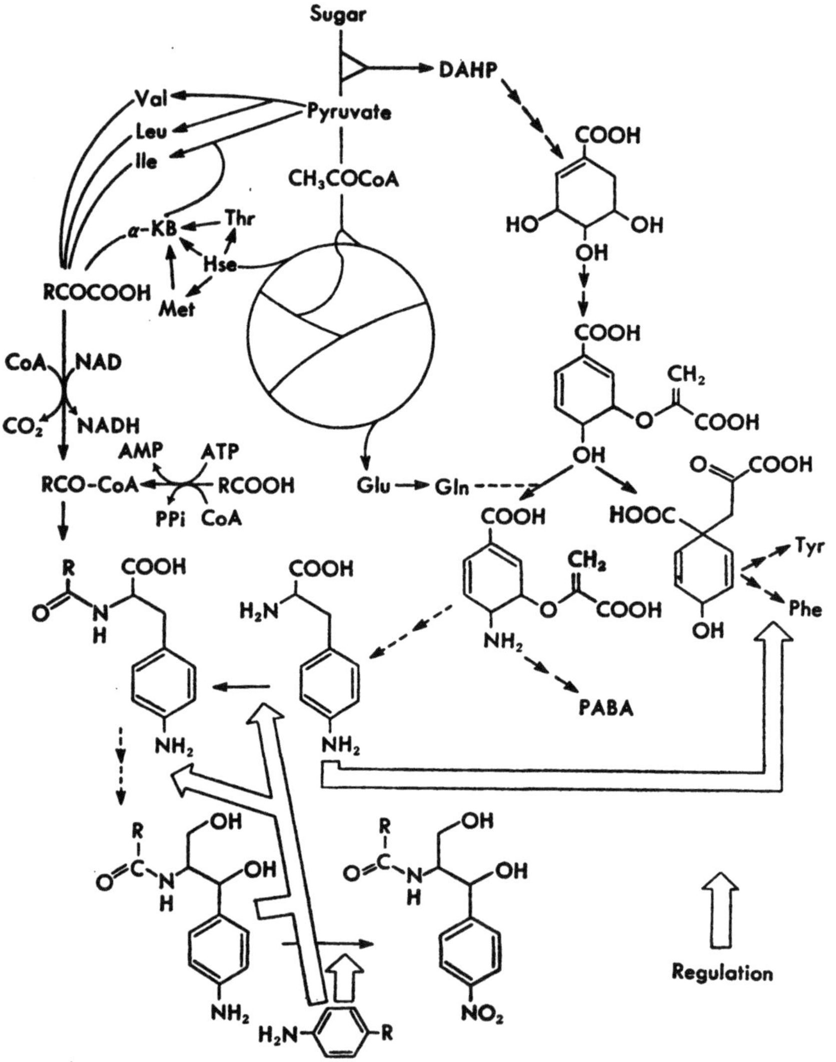

Fig. 9 Synthetic pathways of corynecins and its regulation in *Corynebacterium hydrocarboclastus*[181]

7.5 Miscellaneous

A psychrophile, *Streptomyces phaeochromogenes* No. 351 produced from alkanes about 10 mg l^{-1} of an antibiotic complex SP-351, the active component being characterized as a cyclicpolylactone antibiotic. SP-351 showed a strong antibacterial activity against gram-positive bacteria and acid-fast bacteria[166,182].

Ps. fluorescens KY 4032 accumulated fluopsins C and F in alkane media. These compounds were active against gram-positive and gram-negative bacteria. In the medium supplemented with 80 mg l^{-1} of cupric chloride about 250 mg l^{-1} of fluopsin C, a copper-containing antibiotic, was produced. On the contrary, an iron-containing analogue fluopsin F was predominant (about 200 mg l^{-1}) in the medium containing 3 mg l^{-1} of cupric chloride[183,184].

Str. amylovorus No. P-3355 produced a new antibiotic, P-3355 from alkanes. This antibiotic, whose structure was determined as (5S, 6S)-6-amino-1-formyl-5-hydroxy-1,3-cyclohexadiene, was active against gram-positive and gram-negative bacteria[185].

Pyoluteorin[186] and its derivatives, one of which was identified as 3′-nitropyoluteorin, were produced by *Ps. aeruginosa* in an alkane medium. These compounds showed herbicidal as well as antibacterial activities[187]. As described previously, *Ps. aeruginosa* KY 4025 produced 8.5 g l^{-1} of rhamnonipids from alkanes. These liposaccharides, identified as 2-0-α-L-rhamnopyranosyl-α-L-rhamnopyranosyl-β-hydroxydecanoyl-β-hydroxydecanoate (R-1) and L-α-rhamnopyranosyl-β-hydroxydecanoyl-β-hydroxydecanoate (R-2), were found to have mycoplasmacidal and antiviral activities *in vitro*. R-2, which is considered to be a precursor of R-1, showed a remarkable bactericidal activity against gram-positive bacteria[88].

The productions of brevimycin by *Brevibact. ammoniagenes*[188] and of the methionine antagonist, L-2-amino-4-methoxy-*trans*-3-butenoic acid by *Ps. aeruginosa*[189] were demonstrated in alkane media.

8 Enzymes and Proteins

Recently, applications of not only extracellular but also intracellular enzymes have attracted increasing interests for industrial, analytical and medical purposes. Although various kinds of enzymes are inducible by alkanes, little information is available on the production of enzymes from alkanes. The present authors expect that the metabolic and physiological features of alkane-utilizable microorganisms would favour the production of useful enzymes.

8.1 Protease

Ps. aeruginosa IFO 3455 produced 4×10^{-3} PU per ml of proteinase in an alkane medium supplemented with yeast extract. This value was about one-half of the maximal yield in a glucose medium[190]. Nakao et al. isolated a potent producer of alkaline protease from alkane substrates. This microorganism, *Fusarium sp.* S-19-5, produced 4,000 PU per ml of protease accampanied by 9 DU per ml of α-amylase and 17 U per ml of ribonuclease. A highly proteolytic, kabicidin-resistant mutant

No. 5-128B derived from the strain S-19-5 accumulated 41,000 PU per ml of alkaline protease, 19 DU per ml of α-amylase and 35 U per ml of ribonuclease[140, 191, 192]. The production of the protease on an industrial scale was successfully carried out, 46,000 PU per ml of the enzyme being obtained in a 2,000 l-vessel[193].

8.2 Lipase

About 0.95 U per ml of lipase was produced by *Brettanomyces lambicus* S19Y1 grown on kerosene. Addition of Tween 20 to the kerosene medium was slightly effective, but olive oil inhibited the enzyme formation[194].

8.3 Peroxisomal Enzymes

Conspicuous appearance of peroxisomes (microbodies) was demonstrated by the present authors in alkane-grown yeast cells in harmony with the enhanced level of catalase activity. Such phenomena, however, were not observed in the cells grown on glucose[195, 196]. The peroxisomes isolated from the alkane-grown cells of *C. tropicalis* pK 233 and the oleate-grown cells of *C. lipolytica* NRRL Y-6795 contained various enzymes including catalase, D-amino acid oxidase, uricase, the enzymes participating in fatty acid degradation, the glyoxylate cycle enzymes, *etc.* (Table 6)[197, 198, 199, 200, 201, 202]. Thus, it has been established that the yeast peroxisomes play indispensable roles in the degradation of fatty acids derived from alkanes. and in the synthesis of C_4-compounds from acetyl-CoAs (Fig. 10)[203, 204]. These peroxisomal enzymes seem to be useful for various purposes.

Table 6. Enzymes detected in peroxisomes of alkane-grown yeasts

Catalase, D-amino acid oxidase, uricase
Acyl-CoA synthetase, fatty acid β-oxidation system, carnitine acetyltransferase
Long-chain alcohol dehydrogenase, long-chain aldehyde dehydrogenase
Isocitrate lyase, malate synthase, NADP-linked isocitrate dehydrogenase
NAD-linked glycerophosphate dehydrogenase, glutamate-oxalacetate transaminase

Catalase, which is a marker enzyme of the peroxisomes, was strongly induced in the alkane-grown cells[205]. The catalase levels in the alkane-grown cells of various yeasts were far higher than those in the glucose-grown cells as shown in Table 7[206]. Under the optimal conditions *C. tropicalis* produced 5,000 U per mg cell or 6,200 U per ml of catalase from alkanes of C_{10}–C_{12}[207].

D-Amino acid oxidase would be useful for determination and degradation of D-amino acids, testing optical purity of L-amino acids, and resolution of racemic amino acids. Although this enzyme was found in the peroxisomes, alkanes did not induce the synthesis of the enzyme. *C. tropicalis* produced about 45 U per ml of D-amino acid oxidase in an alkane medium, but the value was about one-half of that in a glucose medium[208].

Uricase (urate oxidase) is a promising enzyme which may be applicable to uric acid determination and to the enzyme therapy of gout. The alkane-utilizing cells of

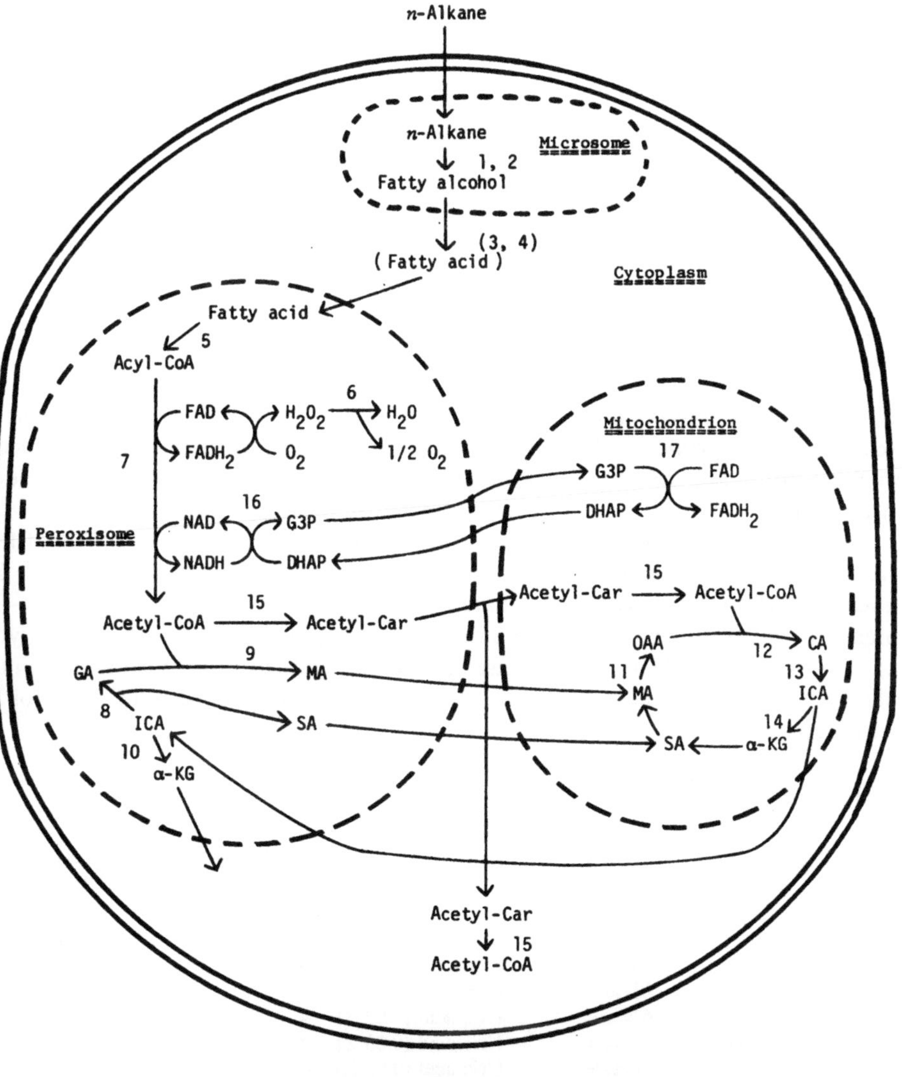

Fig. 10 Presumptive roles of peroxisomes and mitochondria in alkane-assimilating yeats[204).
Enzymes: 1, cytochrome P-450; 2, NADPH-cytochrome P-450 (cytochrome c) reductase; 3, alcohol dehydrogenase; 4, aldehyde dehydrogenase; 5, acyl-CoA synthetase; 6, catalase; 7, β-oxidation system; 8, isocitrate lyase; 9, malate synthase; 10, NADP-linked isocitrate dehydrogenase; 11, malate dehydrogenase; 12, citrate synthase; 13, aconitase; 14, NAD-linked isocitrate dehydrogenase; 15, carnitine acetyltransferase; 16, NAD-linked glycerophosphate dehydrogenase; 17, FAD-linked glycerophosphate dehydrogenase.
CA, citrate; Car, carnitine; DHAP, dihydroxyacetonephosphate; GA, glyoxylate; G3P, glycerol-3-phosphate; ICA, isocitrate; α-KG, α-ketoglutarate; MA, malate; OAA, oxalacetate; SA, succinate

Table 7. Catalase activity of *Candida* yeats at exponential growth phase[203]

Yeast	Relative catalase activity of cells grown on		
	Glucose	Ethanol	Alkane
C. albicans IFO 0587	1	1.5	73
C. guilliermondii IFO 0566	1	1.0	18
C. intermedia NRRL Y-6328-1	1	0.4	24
C. lipolytica NRRL Y-6795	1	0.4	7
C. tropicalis pK 233	1	3.6	53

Catalase activity of glucose-grown cells was taken as a standard

C. tropicalis produced only 2 U per g cell of the enzyme. The resting cells of this yeast, however, produced 91 U per g cell or 140 U per l of uricase in an induction medium composed of alkanes, phosphate buffer, Mg^{2+} and xanthine (Table 8). Alkanes were superior to glucose as substrate for the enzyme production[202].

The fatty acid β-oxidation system, which would be useful for the assays of fatty acids and CoA, was induced exclusively by alkanes or higher fatty acids[198]. The level of NAD-linked glycerophosphate dehydrogenase of *C. tropicalis* was 16-fold higher in the alkane-grown cells than in the glucose-grown cells[201]. The key enzymes of the glyoxylate cycle, isocitrate lyase and malate synthase, were also induced in an alkane medium[209]. These peroxisomal enzymes would also be applicable for the various purposes.

Table 8. Optimization of uricase production by *Candida tropicalis*[203]

Medium	Carbon source	Inducer	Incubation time (h)	Uricase activity (n mol min^{-1} per mg dry cell)
Growth medium	Glucose	None	3	1
	Alkane*	None	5	2
	Alkane	Uric acid (0.03%)	4	8
Induction medium	Glucose	Uric acid (0.03%)	4	21
	Acetate	Uric acid (0.03%)	4	28
	Ethanol	Uric acid (0.03%)	4	12
	Alkane	Uric acid (0.03%)	4	47
	Alkane**	Uric acid (1.5 mM)	6	73
	Alkane**	Xanthine (1.5 mM)	6	91
	Alkane**	Guanine (1.5 mM)	6	82
	Alkane**	None	6	11

* A mixture of C_{10}–C_{13}
** After optimization of incubation conditions

8.4 Protein-like Activator

Several compounds, which stimulate microbial growth on alkanes, were isolated. *C. petrophilum* produced "emulsifying factor" composed of peptide and fatty acid from alkanes[210]. *Ps. aeruginosa* S7B1 also accumulated "protein-like activator" in addition to rhamnolipids in an alkane medium. The isolated activator had molecular weight of 14,300 and amino acid residues of 147. The bacterium produced 100–120 mg l^{-1} of the activator on alkanes of C_{15}–C_{18}[211,212,213].

9 Biomass with Low Content of Odd-chain Fatty Acids

In general yeast cells grown on alkane mixtures contain significant proportions of odd-chain fatty acids as cellular component[214]. The nonnatural cellular fatty acids are derived from alkanes of odd carbon chains *via* intact incorporation and/or elongation of carbon skeleton[215,216]. For the purpose of biomass production from alkanes, it will be desirable to obtain yeast strains not accumulating odd-chain fatty acids even when cultivated on odd-chain alkanes.

Recently, the presence of two distinct acyl-CoA synthetases has been demonstrated in alkane- or fatty acid-grown *C. lipolytica* by Numa and his coworkers[217,218]. The

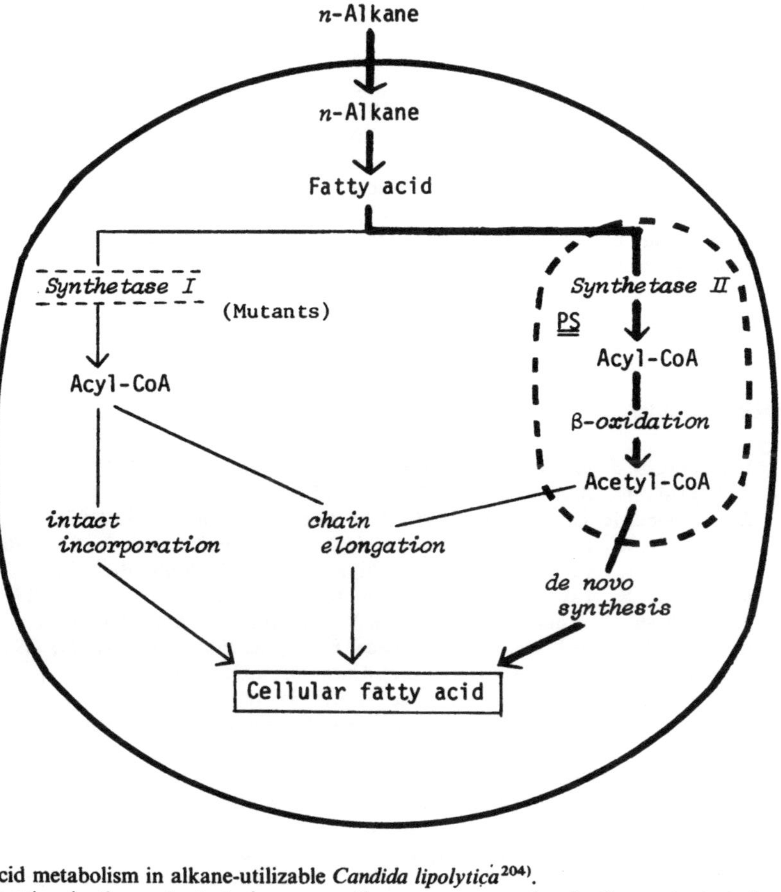

Fig. 11 Fatty acid metabolism in alkane-utilizable *Candida lipolytica*[204].
➡, pathway operating in the mutant strains; →, pathways not operating in the mutant strains

one, "synthetase I", is constitutive and is responsible for the production of acyl-CoA to be utilized for the synthesis of cellular lipids, and the other, "synthetase II", is inducible and is linked to the fatty acid β-oxidation system located in the peroxisomes[199]. Mutant strains lacking synthetase I were isolated by Kamiryo et al.[217]. These strains cannot incorporate exogenous fatty acids as a whole into the cellular lipids, but can degrade them *via* the β-oxidation pathway and synthesize cellular fatty acids *de novo* from acetyl-CoA (Fig. 11). In fact, the proportions of odd-chain cellular fatty acids were drastically reduced in the mutant strains L-5 and L-7 even when the yeast strains were grown on odd-chain alkanes (Table 9)[219], indicating that such the mutants are useful for biomass production.

Table 9. Proportions of odd-chain cellular fatty acids in *Candida lipolytica* wild and mutant strains[219]

Carbon source	Proportion of odd-chain fatty acids (%)	
	Wild strain	Mutant strains
Glucose	3.3	0.7– 0.8
Oleic acid	0.8	0.2– 0.3
n-Undecane	9.2	1.5– 1.7
n-Tridecane	62.0	2.1– 2.4
n-Pentadecane	97.8	8.9–11.4
n-Heptadecane	98.9	11.7–12.4
Alkane mixture 5 (C_{12}–C_{14})	23.5	1.4– 1.8
Alkane mixture 6 (C_{10}–C_{13})	21.9	1.0– 1.1
Alkane mixture 8 (C_{14}–C_{19})	66.8	6.3– 7.0

Cells were harvested at late exponential to early stationary phase

References

1. Abbott, B. J., Gledhill, W. E.: Adv. Appl. Microbiol. *14*, 249 (1971)
2. Yamada, K.: Biotech. Bioeng. *19*, 1563 (1977)
3. Raymond, R. L., Jamison, V. W.: Adv. Appl. Microbiol. *14*, 93 (1971)
4. Yamada, K., Takahashi, J., Kobayashi, K.: Agric. Biol. Chem. *26*, 636 (1962)
5. Yamada, K., Takahashi, J., Kobayashi, K., Imada, Y.: Agric. Biol. Chem. *27*, 390 (1963)
6. Yamada, K., Takahashi, J., Kobayashi, K.: Agric. Biol. Chem. *27*, 773 (1963)
7. Takahashi, J., Kobayashi, K., Imada, Y., Yamada, K.: Appl. Microbiol. *13*, 1 (1965)
8. Imada, Y., Pantskhava, E. C., Yamada, K.: Agric. Biol. Chem. *31*, 245 (1967)
9. Imada, Y., Yamada, K.: Agric. Biol. Chem. *33*, 1326 (1969)
10. Imada, Y., Yamada, K.: Agric. Biol. Chem. *35*, 18 (1971)
11. Shiio, I., Otsuka, S., Ishii, R., Katsuya, N., Iizuka, H.: J. Gen. Appl. Microbiol. *9*, 23 (1963)
12. Otsuka, S., Ishii, R., Shiio, I., Katsuya, N.: J. Gen. Appl. Microbiol. *10*, 179 (1964)
13. Iguchi, T., Hayakawa, S., Takeda, I.: J. Agric. Chem. Soc. Japan *40*, 26 (1966)
14. Iguchi, T., Takeda, I., Seno, S.: Agric. Biol. Chem. *29*, 589 (1965)
15. Shiio, I., Uchio, R.: J. Gen. Appl. Microbiol. *15*, 65 (1969)
16. Yamaguchi, K., Suzuki, T., Kimura, K., Tanaka, K., Kinoshita, S.: Amino Acid. Nucleic Acid *20*, 57 (1969)

17. Suzuki, T., Yamaguchi, K., Tanaka, K.: Agric. Biol. Chem. *35*, 2135 (1971)
18. Hattori, K., Yokoo, S., Imada, O.: J. Ferment. Technol. *52*, 132 (1974)
19. Kobayashi, K., Ikeda, S., Takinami, K., Hirose, Y., Shiro, T.: Agric. Biol. Chem. *35*, 1241 (1971)
20. Nakao, Y., Kanamaru, T., Kikuchi, M., Yamatodani, S.: Agric. Biol. Chem. *37*, 2399 (1973)
21. Kikuchi, M., Kanamaru, T., Nakao, Y.: Agric. Biol. Chem. *37*, 2405 (1973)
22. Nakao, Y., Kikuchi, M., Suzuki, M., Doi, M.: Agric. Biol. Chem. *34*, 1875 (1970)
23. Nakao, Y., Kikuchi, M., Suzuki, M., Doi, M.: Agric. Biol. Chem. *36*, 490 (1972)
24. Kikuchi, M., Suzuki, M., Doi, M., Nakao, Y.: Agric. Biol. Chem. *36*, 809 (1972)
25. Kikuchi, M., Nakao, Y.: Agric. Biol. Chem. *36*, 1135 (1972)
26. Kikuchi, M., Doi, M., Suzuki, M., Nakao, Y.: Agric. Biol. Chem. *36*, 1141 (1972)
27. Kikuchi, M.: J. Agric. Chem. Soc. Japan *49*, R43 (1975)
28. Ishii, R., Otsuka, S., Shiio, I.: J. Gen. Appl. Microbiol. *13*, 217 (1967)
29. Okumura, S.: Petroleum and Microorganisms *7*, 36 (1972)
30. Takayama, K., Abe, S., Kinoshita, S.: Amino Acid. Nucleic Acid *19*, 121 (1969)
31. Kase, H., Nakayama, K.: Agric. Biol. Chem. *37*, 1643 (1973)
32. Hagino, H., Kase, H., Nakayama, K.: J. Agric. Chem. Soc. Japan *48*, 285 (1974)
33. Kase, H., Nakayama, K.: J. Agric. Chem. Soc. Japan *48*, 209 (1974)
34. Oshima, K., Tokoro, Y., Okii, M., Tanaka, K., Kinoshita, S.: Amino Acid. Nucleic Acid *20*, 1 (1969)
35. Uchio, R., Otsuka, S., Shiio, I.: J. Gen. Appl. Microbiol. *13*, 303 (1967)
36. Tanaka, K., Oshima, K., Tokoro, Y., Kinoshita, S.: Amino Acid. Nucleic Acid *18*, 57 (1968)
37. Tokoro, Y., Oshima, K., Tanaka, K., Kinoshita, S.: Amino Acid. Nucleic Acid *19*, 115 (1969)
38. Yamaguchi, K., Tokoro, Y., Oshima, K., Tanaka, K., Suzuki, T.: Agric. Biol. Chem. *37*, 2189 (1973)
39. Tokoro, Y., Oshima, K., Okii, M., Yamaguchi, K., Tanaka, K., Kinoshita, S.: Agric. Biol. Chem. *34*, 1516 (1970)
40. Tomita, F., Suzuki, T., Furuya, A.: Amino Acid. Nucleic Acid *32*, 1 (1975)
41. Tanaka, K., Kimura, K., Suzuki, T., Yamaguchi, K., Kinoshita, S.: J. Ferment. Technol. *47*, 291 (1969)
42. Tsugawa, R., Nakase, T., Kobayashi, T., Yamashita, K., Okumura, S.: Agric. Biol. Chem. *33*, 158 (1969)
43. Tsugawa, R., Okumura, S.: Agric. Biol. Chem. *33*, 676 (1969)
44. Tsugawa, R., Nakase, T., Kobayashi, T., Yamashita, K., Okumura, S.: Agric. Biol. Chem. *33*, 929 (1969)
45. Tabuchi, T., Tanaka, M., Abe, M.: J. Agric. Chem. Soc. Japan *43*, 154 (1969)
46. Abe, M., Tabuchi, T., Tanaka, M.: J. Agric. Chem. Soc. Japan *44*, 493 (1970)
47. Tabuchi, T., Tanaka, M., Tahara, Y., Abe, M.: J. Agric. Chem. Soc. Japan *44*, 562 (1970)
48. Tabuchi, T.: J. Agric. Chem. Soc. Japan *47*, 479 (1973)
49. Tabuchi, T., Tahara, Y., Tanaka, M., Yanagiuchi, S.: J. Agric. Chem. Soc. Japan *47*, 617 (1973)
50. Tabuchi, T., Hara, S.: J. Agric. Chem. Soc. Japan *48*, 417 (1974)
51. Akiyama, S., Suzuki, T., Sumino, Y., Nakao, Y., Fukuda, H.: Agric. Biol. Chem. *36*, 339 (1972)
52. Akiyama, S., Suzuki, T., Sumino, Y., Nakao, Y., Fukuda, H.: Agric. Biol. Chem. *37*, 879 (1973)
53. Akiyama, S., Suzuki, T., Sumino, Y., Nakao, Y., Fukuda, H.: Agric. Biol. Chem. *37*, 885 (1973)
54. Nakanishi, T., Yamamoto, M., Kimura, K., Tanaka, K.: J. Ferment. Technol. *50*, 855 (1972)
55. Hattori, K., Yokoo, S., Imada, O.: J. Ferment. Technol. *52*, 542 (1974)
56. Ikeno, Y., Masuda, M., Tanno, K., Oomori, I., Takahashi, N.: J. Ferment. Technol. *53*, 752 (1975)
57. Nakao, Y.: Petroleum and Microorganisms *9*, 6 (1973)
58. Furukawa, T., Matsuyoshi, T., Minoda, Y., Yamada, K.: J. Ferment. Technol. *55*, 356 (1977)

59. Aiba, S., Matsuoka, M.: Europ. J. Appl. Microbiol. Biotechnol. *5*, 247 (1978)
60. Tabuchi, T., Hara, S.: J. Agric. Chem. Soc. Japan *48*, 543 (1974)
61. Tabuchi, T., Hara, S.: Agric. Biol. Chem. *38*, 1105 (1974)
62. Tabuchi, T., Hara, S.: J. Agric. Chem. Soc. Japan *48*, 549 (1974)
63. Tabuchi, T., Serizawa, N.: Agric. Biol. Chem. *39*, 1049 (1975)
64. Tabuchi, T., Serizawa, N., Ohmono, S.: Agric. Biol. Chem. *38*, 2565 (1974)
65. Tabuchi, T.: J. Agric. Chem. Soc. Japan *52*, R1 (1978)
66. Yamada, K., Furukawa, T., Nakahara, T.: Agric. Biol. Chem. *34*, 670 (1970)
67. Furukawa, T., Nakahara, T., Yamada, K.: Agric. Biol. Chem. *34*, 1402 (1970)
68. Furukawa, T., Nakahara, T., Yamada, K.: Agric. Biol. Chem. *34*, 1833 (1970)
69. Furukawa, T., de Miranda, L. R., Matsuyoshi, T.: J. Ferment. Technol. *56*, 546 (1978)
70. Sato, S., Nakahara, T., Minoda, Y.: Agric. Biol. Chem. *41*, 967 (1977)
71. Sato, M., Nakahara, T., Yamada, K.: Agric. Biol. Chem. *36*, 1745 (1972)
72. Sato, M., Nakahara, T., Yamada, K.: Agric. Biol. Chem. *36*, 1969 (1972)
73. Sato, M., Nakahara, T., Yamada, K.: Agric. Biol. Chem. *36*, 2025 (1972)
74. Meiki, S., Tabuchi, T.: J. Agric. Chem. Soc. Japan *49*, 325 (1975)
75. Uchio, R., Shiio, I.: Petroleum and Microorganisms *11*, 14 (1974)
76. Ogino, S., Yano, K., Tamura, G., Arima, K.: Agric. Biol. Chem. *29*, 1009 (1965)
77. Iizuka, H., Iida M., Unami, Y.: J. Gen. Appl. Microbiol. *12*, 119 (1966)
78. Okuhara, M., Kubochi, Y., Harada, T.: Agric. Biol. Chem. *35*, 1376 (1971)
79. Ogata, K., Kaneyuki, H., Kato, N., Tani, Y., Yamada, H.: J. Ferment. Technol. *51*, 227 (1973)
80. Shiio, I., Uchio, R.: Agric. Biol. Chem. *35*, 2033 (1971)
81. Uchio, R., Shiio, I.: Agric. Biol. Chem. *36*, 426 (1972)
82. Uchio, R., Shiio, I.: Agric. Biol. Chem. *36*, 1169 (1972)
83. Uchio, R., Shiio, I.: Agric. Biol. Chem. *36*, 1389 (1972)
84. Uchio, R.: Petroleum and Microorganisms *20*, 13 (1978)
85. Suzuki, T., Tanaka, K., Kinoshita, S.: Agric. Biol. Chem. *33*, 190 (1969)
86. Suzuki, T., Tanaka, K., Matsubara, I., Kinoshita, S.: Agric. Biol. Chem. *33*, 1619 (1969)
87. Hisatsuka, K., Nakahara, T., Sano, N., Yamada, K.: Agric. Biol. Chem. *35*, 686 (1971)
88. Itoh, S., Honda, H., Tomita, F., Suzuki, T.: J. Antibiotics *24*, 855 (1971)
89. Kanamaru, T., Yamatodani, S.: Agric. Biol. Chem. *33*, 1521 (1969)
90. Tabuchi, T., Hara, S.: J. Agric. Chem. Soc. Japan *47*, 485 (1973)
91. Hattori, K., Suzuki, T.: Agric. Biol. Chem. *38*, 581 (1974)
92. Hattori, K., Suzuki, T.: Agric. Biol. Chem. *38*, 1203 (1974)
93. Hattori, K., Suzuki, T.: Agric. Biol. Chem. *38*, 2419 (1974)
94. Hattori, K., Suzuki, T.: Agric. Biol. Chem. *39*, 57 (1975)
95. Hattori, K., Suzuki, T.: Agric. Biol. Chem. *38*, 1875 (1974)
96. Iguchi, T., Takeda, I.: Agric. Biol. Chem. *30*, 709 (1966)
97. Iguchi, T., Watanabe, T., Takeda, I.: Agric. Biol. Chem. *31*, 569 (1967)
98. Iguchi, T., Watanabe, T., Takeda, I.: Agric. Biol. Chem. *31*, 574 (1967)
99. Iguchi, T., Kodaira, R., Takeda, I.: Agric. Biol. Chem. *31*, 885 (1967)
100. Kawamoto, I., Nara, T., Misawa, M., Kinoshita, S.: Agric. Biol. Chem. *34*, 1142 (1970)
101. Takeda, I., Mizuno, M., Shimojima, K.: Amino Acid. Nucleic Acid *22*, 81 (1970)
102. Tomita, F., Suzuki, T.: Agric. Biol. Chem. *38*, 71 (1974)
103. Watanabe, S., Takeda, I.: Agric. Biol. Chem. *36*, 2265 (1972)
104. Tomita, F., Suzuki, T.: Agric. Biol. Chem. *36*, 133 (1972)
105. Tomita, F., Nakanishi, T., Suzuki, T.: Agric. Biol. Chem. *38*, 293 (1974)
106. Asai, T., Kuno, M., Nakao, Y., Kono, T., Fukuda, H.: J. Ferment. Technol. *56*, 75 (1978)
107. Nishio, N., Kamikubo, T.: J. Ferment. Technol. *48*, 1 (1970)
108. Nishio, N., Kamikubo, T.: Agric. Biol. Chem. *35*, 485 (1971)
109. Teranishi, Y., Shimizu, S., Tanaka, A., Fukui, S.: J. Ferment. Technol. *49*, 213 (1971)
110. Nishimura, Y., Iizuka, H.: J. Agric. Chem. Soc. Japan *46*, 639 (1972)
111. Nishimura, Y., Iizuka, H.: J. Gen. Appl. Microbiol. *22*, 79 (1976)
112. Nishimura, Y., Iizuka, H.: Z. Allg. Mikrobiol. *12*, 231 (1972)
113. Tanaka, A., Fukui, S.: J. Ferment. Technol. *45*, 611 (1967)

114. Tanaka, A., Ohishi, N., Fukui, S.: J. Ferment. Technol. *45*, 617 (1967)
115. Nishio, N., Sakai, K., Fujita, K., Kamikubo, T.: Agric. Biol. Chem. *37*, 553 (1973)
116. Fujii, K., Shimizu, S., Fukui, S.: J. Ferment. Technol. *44*, 185 (1966)
117. Fukui, S., Shimizu, S., Fujii, K.: J. Ferment. Technol. *45*, 530 (1967)
118. Morikawa, H., Kamikubo, T.: J. Ferment. Technol. *47*, 470 (1969)
119. Morikawa, H., Hayashi, M., Kamikubo, T.: J. Ferment. Technol. *49*, 803 (1971)
120. Nishio, N., Ueda, M., Omae, Y., Hayashi, M., Kamikubo, T.: Agric. Biol. Chem. *40*, 2037 (1976)
121. Tsuboi, T., Sekijo, C., Shoji, O.: Agric. Biol. Chem. *30*, 1238 (1966)
122. Tsuboi, T., Sekijo, C., Shoji, O.: Agric. Biol. Chem. *30*, 1243 (1966)
123. Tsuboi, T., Sekijo, C., Yoshimura, Y., Shoji, O.: Agric. Biol. Chem. *31*, 1135 (1967)
124. Ogino, S., Fujimoto, S., Aoki, Y.: Agric. Biol. Chem. *38*, 275 (1974)
125. Ogino, S., Fujimoto, S., Aoki, Y.: Agric. Biol. Chem. *38*, 707 (1974)
126. Kuno, M., Kikuchi, M., Nakao, Y.: Agric. Biol. Chem. *37*, 313 (1973)
127. Tanaka, A., Maki, H., Fukui, S.: J. Ferment. Technol. *45*, 1163 (1967)
128. Tanaka, A., Fukui, S.: J. Ferment. Technol. *48*, 137 (1970)
129. Tanaka, A., Nagasaki, T., Fukui, S.: J. Ferment. Technol. *46*, 984 (1968)
130. Tanaka, A., Fujii, K., Fukui, S.: J. Ferment. Technol. *47*, 297 (1969)
131. Fujii, K., Tanaka, A., Fukui, S.: Arch. Microbiol. *97*, 259 (1974)
132. Takeda, I., Hayakawa, S.: Agric. Biol. Chem. *32*, 873 (1968)
133. Takeda, I.: Agric. Biol. Chem. *33*, 122 (1969)
134. McNutt, W. S., Takeda, I.: Biochemistry *8*, 1370 (1969)
135. Takeda, I.: Amino Acid. Nucleic Acid *22*, 133 (1970)
136. Shirota, S., Hayakawa, S., Watanabe, S., Iguchi, T., Takeda, I.: J. Ferment. Technol. *48*, 747 (1970)
137. Tanaka, A., Yamada, R., Shimizu, S., Fukui, S.: J. Ferment. Technol. *49*, 792 (1971)
138. Lin, H. T., Iida, M., Iizuka, H.: J. Ferment. Technol. *49*, 206 (1971)
139. Lin, H. T., Iida, M., Iizuka, H.: J. Ferment. Technol. *49*, 771 (1971)
140. Nakao, Y., Suzuki, M., Kuno, M., Maejima, K.: Agric. Biol. Chem. *37*, 1223 (1973)
141. Tanaka, A., Nagasaki, T., Inagawa, M., Fukui, S.: J. Ferment. Technol. *46*, 468 (1968)
142. Tanaka, A., Nagasaki, T., Fukui, S.: J. Ferment. Technol. *46*, 477 (1968)
143. Tanaka, A., Nagasaki, T., Fukui, S., Murakami, S.: J. Ferment. Technol. *47*, 739 (1969)
144. Tanaka, A., Kato, K., Fukui, S.: J. Ferment. Technol. *49*, 778 (1971)
145. Sakurai, H., Kato, K., Sakai, T., Masuda, Y., Kuriyama, T.: Bull. Chem. Soc. Japan *44*, 481 (1971)
146. Hsieh, L. K., Lee, T. C., Chichester, C. O., Simpson, K. L.: J. Bacteriol. *118*, 385 (1974)
147. Iizuka, H., Nishimura, Y.: J. Gen. Appl. Microbiol. *15*, 127 (1969)
148. Takeda, I., Tsuchimoto, N., Miwa, S.: Appl. Microbiol. *16*, 1806 (1968)
149. Shimizu, S., Tanaka, A., Fukui, S.: J. Ferment. Technol. *47*, 542 (1969)
150. Shimizu, S., Tanaka, A., Fukui, S.: J. Ferment. Technol. *47*, 551 (1969)
151. Shimizu, S., Tanaka, A., Fukui, S.: J. Ferment. Technol. *48*, 533 (1970)
152. Shimizu, S., Tanaka, A., Fukui, S.: J. Ferment. Technol. *48*, 542 (1970)
153. Shimizu, S., Tanaka, A., Fukui, S.: J. Ferment. Technol. *48*, 549 (1970)
154. Tanaka, A., Shimizu, S., Fukui, S.: J. Ferment. Technol. *52*, 925 (1974)
155. Fukui, S., Tanaka, A.: Proc. 8th World Petroleum Congress, Vol. 5, p. 157, 1971
156. Morimoto, H., Imada, I., Watanabe, M., Nakao, Y.: Liebigs Ann. Chem. *729*, 158 (1969)
157. Lee, E. G. H., Walden, C. C.: Appl. Microbiol. *17*, 520 (1969)
158. Tanaka, A., Taira, T., Teranishi, Y., Fukui, S.: J. Ferment. Technol. *50*, 662 (1972)
159. Higashihara, T., Sato, A.: Agric. Biol. Chem. *33*, 1802 (1969)
160. Higashihara, T., Sato, A.: J. Ferment. Technol. *48*, 73 (1970)
161. Ogata, K., Minami, S., Tani, Y.: J. Ferment. Technol. *49*, 925 (1971)
162. Suzuki, T., Uno, K., Deguchi, T.: Agric. Biol. Chem. *35*, 92 (1971)
163. Kitano, K., Kintaka, K., Suzuki, S., Katamoto, K., Nara, K., Nakao, Y.: J. Ferment. Technol. *52*, 785 (1974)
164. Kitano, K., Kintaka, K., Suzuki, S., Katamoto, K., Nara, K., Nakao, Y.: J. Ferment. Technol. *54*, 683 (1976)
165. Ogata, K., Yoshida, N., Ohsugi, M., Tani, Y.: Agric. Biol. Chem. *35*, 79 (1971)

166. Yoshida, N., Tani, Y., Ogata, K.: J. Antibiotics *25*, 546 (1972)
167. Yoshida, N., Tani, Y., Ogata, K.: J. Antibiotics *25*, 653 (1972)
168. Yoshida, N., Hanochi, H., Hachiya, Y., Tani, Y., Ogata, K.: J. Antibiotics *26*, 60 (1973)
169. Yoshida, N., Hayashi, S., Tani, Y., Ogata, K.: J. Antibiotics *27*, 128 (1974)
170. Suzuki, T., Honda, H., Katsumata, R.: Agric. Biol. Chem. *36*, 2223 (1972)
171. Shirahata, K., Hayashi, T., Deguchi, T., Suzuki, T., Matsubara, I.: Agric. Biol. Chem. *36*, 2229 (1972)
172. Tomita, F., Nakano, H., Honda, H., Suzuki, T.: Agric. Biol. Chem. *38*, 2183 (1974)
173. Nakano, H., Tomita, F., Suzuki, T.: Agric. Biol. Chem. *38*, 2471 (1974)
174. Tomita, F., Nakano, H., Suzuki, T.: Agric. Biol. Chem. *38*, 1673 (1974)
175. Suzuki, T., Tomita, F., Nakano, H.: Agric. Biol. Chem. *38*, 2477 (1974)
176. Nakano, H., Tomita, F., Yamaguchi, K., Nagashima, M., Suzuki, T.: Biotechnol. Bioeng. *19*, 1009 (1977)
177. Nakano, H., Tomita, F., Suzuki, T.: Agric. Biol. Chem. *38*, 2505 (1974)
178. Nakano, H., Tomita, F., Suzuki, T.: Agric. Biol. Chem. *40*, 207 (1976)
179. Nakano, H., Tomita, F., Suzuki, T.: Agric. Biol. Chem. *40*, 331 (1976)
180. Nakano, H.: Petroleum and Microorganisms *20*, 4 (1978)
181. Tomita, F.: Kagaku to Seibutsu *16*, 480 (1978)
182. Yoshida, N., Hachiya, Y., Ueda, K., Tani, Y., Ogata, K.: Agric. Biol. Chem. *37*, 661 (1973)
183. Itoh, S., Inuzuka, K., Suzuki, T.: J. Antibiotics *23*, 542 (1970)
184. Shirahata, K., Dguchi, T., Hayashi, T., Matsubara, I., Suzuki, T.: J. Antibiotics *23*, 546 (1970)
185. Sumino, Y., Akiyama, S., Haibara, K., Asai, M., Mizuno, K.: J. Antibiotics *29*, 479 (1976)
186. Birch, A. J., Hodge, P., Rickards, R. W., Takeda, R., Watson, T. R.: J. Chem. Soc. *1964*, 2641
187. Ohmori, T., Hagiwara, S., Ueda, A., Minoda, Y., Yamada, K.: Agric. Biol. Chem. *42*, 2031 (1978)
188. Murdia U. S., Tamhane, D. V.: J. Ferment. Technol. *52*, 598 (1974)
189. Sahm, U., Knobloch, G., Wagner, F.: J. Antibiotics *26*, 389 (1973)
190. Morihara, K.: Appl. Microbiol. *13*, 793 (1965)
191. Suzuki, M., Kuno, M., Maejima, K., Nakao, Y.: Agric. Biol. Chem. *38*, 135 (1974)
192. Suzuki, M., Kuno, M., Asai, T., Nakao, Y.: Agric. Biol. Chem. *38*, 657 (1974)
193. Asai, T., Nakao, Y., Kono, T., Fukuda, H.: J. Ferment. Technol. *56*, 84 (1978)
194. Takahashi, J., Imada, Y., Yamada, K.: Agric. Biol. Chem. *27*, 396 (1963)
195. Osumi, M., Miwa, N., Teranishi, Y., Tanaka, A., Fukui, S.: Arch. Microbiol. *99*, 181 (1974)
196. Osumi, M., Fukuzumi, F., Teranishi, Y., Tanaka, A., Fukui, S.: Arch. Microbiol. *103*, 1 (1975)
197. Kawamoto, S., Tanaka, A., Yamamura, M., Teranishi, Y., Fukui, S., Osumi, M.: Arch. Microbiol. *112*, 1 (1977)
198. Kawamoto, S., Nozaki, C., Tanaka, A., Fukui, S.: Eur. J. Biochem. *83*, 609 (1978)
199. Mishina, M., Kamiryo, T., Tashiro, S., Hagihara, T., Tanaka, A., Fukui, S., Numa, S.: Eur. J. Biochem. *89*, 321 (1978)
200. Kawamoto, S., Ueda, M., Nozaki, C., Yamamura, M., Tanaka, A., Fukui, S.: FEBS Lett. *96*, 37 (1978)
201. Kawamoto, S., Yamada, T., Tanaka, A., Fukui, S.: FEBS Lett. *97* 253 (1979)
202. Tanaka, A., Yamamura, M., Kawamoto, S., Fukui, S.: Appl. Environ. Microbiol. *34*, 342 (1977)
203. Fukui, S., Tanaka, A.: J. Appl. Biochem. *1*, 171 (1979)
204. Fukui, S., Tanaka, A.: Trends in Biochem. Sci. *4*, 246 (1979)
205. Teranishi, Y., Kawamoto, S., Tanaka, A., Osumi, M., Fukui, S.: Agric. Biol. Chem. *38*, 1221 (1974)
206. Teranishi, Y., Tanaka, A., Osumi, M., Fukui, S.: Agric. Biol. Chem. *38*, 1213 (1974)
207. Tanaka, A., Takahashi, R., Kawamoto, S., Fukui, S.: J. Ferment. Technol. *54*, 850 (1976)
208. Kawamoto, S., Kobayashi, M., Tanaka, A., Fukui, S.: J. Ferment. Technol. *55*, 13 (1977)
209. Nabeshima, S., Tanaka, A., Fukui, S.: Agric. Biol. Chem. *41*, 275 (1977)
210. Iguchi, T., Takeda, I., Ohsawa, H.: Agric. Biol. Chem. *33*, 1657 (1969)
211. Hisatsuka, K., Nakahara, T., Yamada, K.: Agric. Biol. Chem. *36*, 1361 (1972)
212. Hisatsuka, K., Nakahara, T., Minoda, Y., Yamada, K.: Agric. Biol. Chem. *39*, 999 (1975)

213. Hisatsuka, K., Nakahara, T., Minoda, Y., Yamada, K.: Agric. Biol. Chem. *41*, 445 (1977)
214. Rattray, J. M. B., Schibeci, A., Kidby, D. K.: Bacteriol. Rev. *39*, 197 (1975)
215. Mishina, M., Yanagawa, S., Tanaka, A., Fukui, S.: Agric. Biol. Chem. *37*, 863 (1973)
216. Tanaka, A., Hagihara, T., Nishikawa, Y., Mishina, M., Fukui, S.: Europ. J. Appl. Microbiol. *3*, 115 (1976)
217. Kamiryo, T., Mishina, M., Tashiro, S., Numa, S.: Proc. Natl. Acad. Sci. USA *74*, 4947 (1977)
218. Mishina, M., Kamiryo, T., Tashiro, S., Numa, S.: Eur. J. Biochem. *82* 347 (1978)
219. Tanaka, A., Hagihara, T., Kamiryo, T., Mishina, M., Tashiro, S., Numa, S., Fukui, S.: Europ. J. Appl. Microbiol. Biotechnol. *5*, 79 (1978)

Microbial Reactions in Prostaglandin Chemistry

J. Jiu, Ph. D.
G. D. Searle and Company, Research and Development Division,
Post Office Box 5110, Chicago, Illinois 60680, U.S.A.

1 Introduction ... 38
2 Applications of Microbial Transformation in Prostaglandin Synthesis ... 41
2.1 Direct Microbial Synthesis of Prostaglandins ... 41
2.2 Microbial Reactions in the Total Synthesis ... 43
2.3 Microbial Transformations of Prostaglandins ... 51
3 Classification by Microbial Reaction Types ... 55
3.1 Microbial Cyclizations ... 55
3.2 Microbial Reductions ... 56
3.2.1 Microbial Reductions of Double Bonds ... 56
3.2.2 Microbial Reductions of Ketones ... 56
3.3 Microbial Oxidations ... 58
3.3.1 Microbial Oxidation of Alcohol to Ketone ... 58
3.3.2 Microbial Hydroxylations ... 58
3.3.3 Degradative Microbial Oxidations ... 58
3.4 Microbial Hydrolysis ... 59
3.4.1 Optical Resolutions via Microbial Hydrolysis ... 59
3.4.2 Microbial Hydrolysis of Esters ... 60
4 Conclusion ... 60
5 Acknowlegment ... 61
6 References ... 61

Microbial reactions are used in the preparation of prostaglandins and prostaglandin derivatives. Although the applications of microbial reactions in prostaglandin chemistry is in its earlier stages, many microbial reactions have already been elucidated. Prostaglandins and prostaglandin-like substances have been prepared microbiologically from carbon and nitrogen sources. Microbial reactions have been used as an integral part of total synthetic schemes. Particularly, microbial reactions have been used to prepare optically active intermediates or products. Various prostaglandin derivatives have been prepared by microbial transformation for structure-activity correlations.

The various microbial reactions elucidated are: microbial reductions of ketones and double bonds, microbial hydroxylations, microbial oxidations and microbial hydrolysis of esters. The feasibility of microbial reactions in prostaglandin chemistry has been documented. The potential demands for microbial reactions in prostaglandin chemistry is enhanced as the demands for optically pure prostaglandins increases.

1 Introduction

During the 1930s, the prostaglandins quietly came on the scientific scene. Kurzrok and Lieb[1] noted that fresh human seminal fluid contained factors which could stimulate the human uterus tissue to either relax or contract, depending on the preconditioning of the uterus strip. Independently Goldblatt[2] and von Euler[3] demonstrated that extracts of human seminal plasma or sheep vesicular glands could stimulate smooth muscle preparations and lower the blood pressures of various experimental animals. This was the period of great advances in pharmacological analysis and the use of specific inhibitors to differentiate between various naturally occuring biologically active compounds. Von Euler demonstrated that the active factors were different from such known biologically active compounds as adrenaline, histamine, acetylcholine, or substance P[4,5]. Von Euler named the active factors "prostaglandins", since these new factors were found in highest concentration in the prostate glands.

The active factors were demonstrated to be lipid-like substances. Initially, progress in prostaglandin research was slow. This was attributed to several problems. One of the problems was the scarcity of raw materials. In addition the active factors were present in submilligram quantities. However, the major impediment was the lack of good methodologies for the isolation, purification and identification of the lipid-like substances. Without the actual individual active factors, no conclusion could be drawn as to the nature and specificity of the active factors.

After the war, Bergström addressed the task of isolating, purifying and identifying the active factors. During this process, elegant separation methods were developed for the isolation and purification of prostaglandins. Also, microanalytical methods were developed for the identification of the active factors. By 1957 the first crystalline product was isolated[6]. This was subsequently identified to be $PGF_{2\alpha}$ [7]. This was followed by the isolation and identification of PGE_2 [8]. By the mid-1960s, a whole family of closely related prostaglandins were isolated and identified[9,10,11].

The isolation of the individual prostaglandins and the elucidation of their structures resulted in an exponential increase in prostaglandin research. In a short time, prostaglandin research rose from a laboratory curiosity to one of the major research areas.

The prostaglandins were shown to be ubiquitous, having been detected in low concentrations in practically all tissues investigated[12]. They are biosynthesized from essential fatty acids[13,14], and are readily metabolized. The prostaglandins exhibited a wide, and diverse range of pharmacological actions[13,14,15]. They are potent compounds and elicit their biological effects at exceedingly low concentrations. The range of biological effects suggest that they may be local modulators of cell functions. The prostaglandins are involved in the regulation or modulation of endocrine, reproductive, nervous, digestive, hemostatic, respiratory, cardiovascular, renal lipid and carbohydrate systems. Their mode of actions has not been clearly defined, though they may be involved in the secondary messenger systems. Possessing such broad and diverse potent biological properties, the prostaglandins have great clinical potentials.

These findings resulted in an increase in demands for the prostaglandins. Compounds were used to evaluate the biological parameters of the individual prostaglandins. Raw materials were used to synthesize prostaglandin derivatives in order to define their structure-activity relationships. The alternative routes used to meet these demands were scale-up biosynthesis[16,17] and total chemical synthesis[18,19,20]. Both alternatives were capable of producing new and novel prostaglandin derivatives. The scale-up biosynthesis provided much materials used to evaluate the biochemical and biological properties of the prostaglandins. The phenomenal growth in the number of chemical syntheses of prostaglandins, resulted in the availability of several prostaglandin intermediates commercially[18].

Most of the total synthetic schemes produced racemic prostaglandins. Many of the unnatural isomers themselves, exhibited biological properties unique and different from the natural isomers. The separations of the enantiomers are tedious and expensive processes. Several alternatives were used to prepare optically active prostaglandins, such as: the use of optically active starting materials; the use of stereospecific chemical reactions; and the use of enzymatic reactions. The application of enzymatic reactions is one of the more elegant methods for the synthesis of optically active compounds. This report is a survey of the applications of microbial processes in the synthesis of prostaglandins, or prostaglandin derivatives.

The prostaglandins are a family of natural, cyclic fatty acids, widely distributed in mammalian systems. The more common ones are known as the primary prostaglandins. Their structures are shown in Fig. 1. The numbering system used for the prostaglandins is shown in Fig. 2. Starting with the carboxylic acid carbon, the compound is numbered sequentially around the chain, ending with the terminal methyl group as C-20. At times, this prostaglandin numbering system will be used to designate a carbon atom in a prostaglandin synthon or a partial structure.

Two nomenclature systems are used to denote the conformation at a given carbon atom. The symbols α (.........., broken bond) and β (———, solid bond) are used

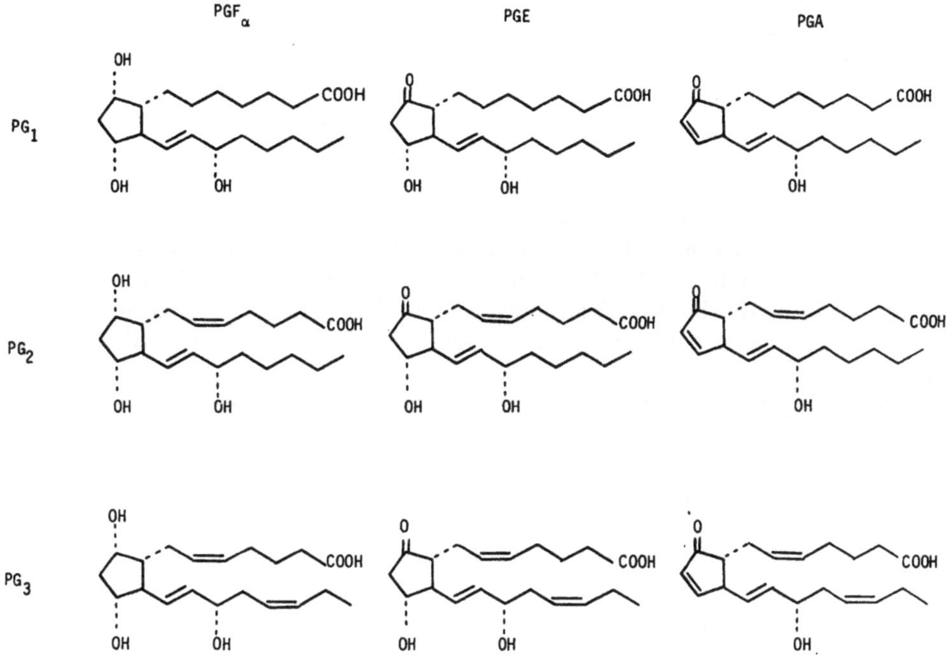

Fig. 1 Structures of primary prostaglandins

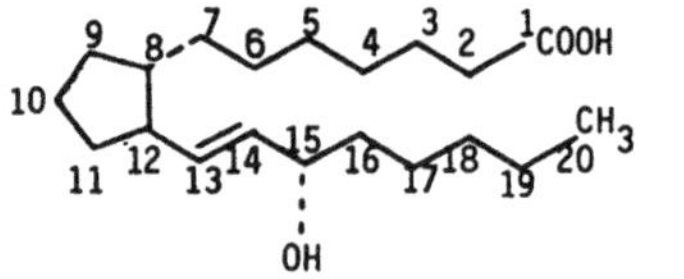

Fig. 2 Numbering system used

to denote the projection of a bond relative to the plane of the page (Fig. 3). The Cahn-Ingold-Prelog symbols R and S are used to denote the chirality of a bond at an asymmetric carbon. Additionally, the symbols, α and ω, are used to designate the two aliphatic side chains attached to the cyclopentane ring.

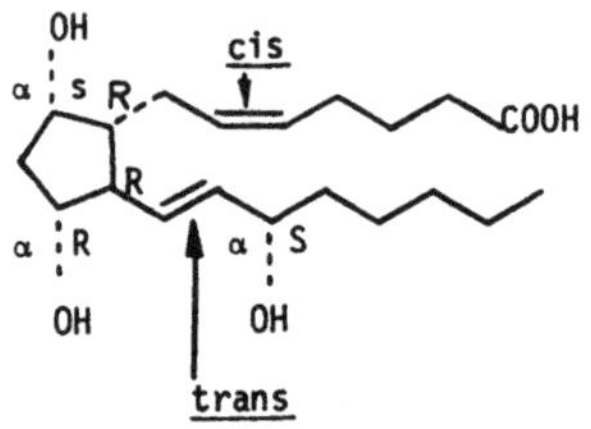

Fig. 3 Projection symbols

Some of the salient features of the primary prostaglandins (Fig. 2) are that they are C-20 fatty acids, with a cyclopentane ring, a *trans* double bond at C-13, 14, and an α-hydroxy group at C-15 with S chirality. These are the common features of all the primary prostaglandins.

The PGE series (Fig. 1, middle column) have in addition to the common features, a ketone group at C-9, and an α-hydroxyl group at C-11 on the cyclopentane ring at R chirality. The PGF series have hydroxyl groups both at C-9 and C-11. The PGF series is divided into 2 subgroups. When the C-9 alcohol is an α-hydroxyl group (S chirality), then the compounds are in the PGFα subseries (Fig. 1, first column). When the C-9 alcohol is a β-hydroxy group (R chirality), then the compounds are in the PGFβ subseries. In both subseries, the C-11 alcohol is an α-hydroxy group (R chirality). The elimination fo the C-11α-hydroxyl group to form the C-10, 11 double bond yields the compounds of the PGA series (Fig. 1, last column).

The subscripts 1, 2, and 3 denote the number of double bonds present in the prostaglandin molecule. The prostaglandin with the subscript 1, (Fig. 1, top row) denotes the presence of a single double bond, a *trans* double bond at C-13, 14. The subscript 2 denotes the presence of a second double bond, a *cis* double bond at C-5, 6, in addition to the *trans* double bond at C-13, 14 (middle row). The prostaglandins with subscript 3 have a third double bond, a *cis* double bond at C-17, 18 (bottom row).

The potent biological properties of the primary prostaglandins and the wide variety of physiological responses they elicit created much interest in the total synthesis of these compounds. However, the complexity of the prostaglandin structures presented a synthetic challenge. Using $PGF_{2\alpha}$ as the prototype (Fig. 3), $PGF_{2\alpha}$ has 6 functionalized centers: a carboxylic acid group at C-1; a *cis* double bond at C-5, 6; an α-hydroxyl group at C-9; an α-hydroxyl group at C-11; a *trans* double bond at C-13, 14; and an α-hydroxy group at C-15. In addition, the molecule has 5 asymmetric centers: the attachment of the α-aliphatic side chain to the cyclopentane ring at C-8 with R chirality; the C-9 α-hydroxyl group with S chirality; the C-11α-hydroxy group with R chirality; the attachment of the ω-aliphatic side chain to the cyclopentane ring at C-12 with R chirality; and the C-15α-hydroxyl group with S chirality.

The complexity of the primary prostaglandin structures, especially the numerous asymmetric centers, are reminiscent of the similar type of problems encountered in the synthesis of steroids. Based on the successful application of microbial transformations in the synthesis of steroids, it appears that microbial processes, especially microbial transformations, would be a useful tool for the preparation of prostaglandins or prostaglandin derivatives.

2 Applications of Microbial Transformation in Prostaglandin Synthesis

Various groups have used microbial reactions for the preparation of prostaglandins or prostaglandin derivatives. Some have used the microbial processes to mimic the biosynthesis of prostaglandins. These studies more or less are concerned with the direct microbial synthesis of prostaglandins from C and N sources. Others have used microbial reactions in place of chemical reactions in selective steps in a total synthetic scheme. In these studies, microbial transformations were used as a synthetic tool. Many of the total synthetic schemes have steps which can be carried out by microbial processes. Generally, these steps are associated with the introduction of an asymmetric center. A third group has used microbial reactions for the preparation of new and novel prostaglandin derivatives. In these studies microbial transformations were used to convert a intact prostaglandin molecule to some new and/or novel derivatives.

2.1 Direct Microbial Synthesis of Prostaglandins

Several groups have studied the use of microbial processes for the direct preparation of prostaglandins or prostaglandin-like substances from various carbon and nitrogen sources. In these studies, the prostaglandins are assumed to be prepared microbiologically in a manner similar to the mammalian biosynthetic pathway.

Skarnes and Howard described a process for the microbial preparation of prostaglandin and/or prostaglandin-like compounds[21)]. *Pseudomonas aeruginosa* was grown in media containing carbon and nitrogen sources that were free of any fatty acids. The yields of prostaglandin substances were at the levels of mg ml^{-1}. The addition of arachidonic acid to the culture medium reduced the amount of prostaglandins produced. This ability to produce prostaglandins apparently is unique to *Pseudomonas aeruginosa*. None of the other bacterial species studied produced prostaglandin or prostaglandin-like substances.

Bobylev et al.[22], and Gandel et al.[23], described processes for the "fermentative" conversion of unsaturated fatty acids to prostaglandins. It is not clear from the abstracted articles whether the "fermentative" reactions were biochemical or microbiological processes.

Beal, Fonken and Pike[24] described a process for the microbiological conversion of unsaturated fatty acids to prostaglandins and prostaglandin analogs. The process in its broadest aspect comprises subjecting an unsaturated fatty acid having a 1,4-diene system to the oxygenating activity of a species of the subphylum Thallophyta. Polyunsaturated fatty acids, in general, and arachidonic acid in particular, where subjected to any of the 125 cultures in an appropriate media to produce a mixture of PGE and PGF prostaglandins.

The preceeding three studies described microbiological processes which convert C and/or N sources all the way the prostaglandins. The next group of studies are more specific, in that the microbial processes were used just for the microbial preparation of the fatty acid intermediates. The subsequent conversions of the fatty acids to prostaglandins were achieved by biochemical processes.

Watanabe and Ono[25] described the microbial production of arachidonic acid by culturing *Euglena viridis* in a polypeptone and yeast medium. The harvested alga contained 6.3% arachidonic acid (dry weight).

Iizuka et al.[26] described processes that enhanced the microbiological production of arachidonic acid. *Penicillium*, *Cladosporium*, *Mucor*, *Fusarium*, *Hormodendrum*, *Aspergillus* or *Rhodotorula* species were first cultured in a carbohydrate medium. The culture was then mixed with a hydrocarbon, such as kerosene, or a fatty acid, such as linoleic acid, and further cultured. The second culturing, in the presence of a hydrocarbon or a fatty acid increased the amounts of arachidonic acid produced by 2.3 to 3.4 fold.

Using *Ophiobolus graminis*, Sih et al.[27] first microbiologically hydroxylated arachidonic acid to 18ξ-hydroxyarachidonic acid in 20% yield (Fig. 4, *2*), and 19ξ-

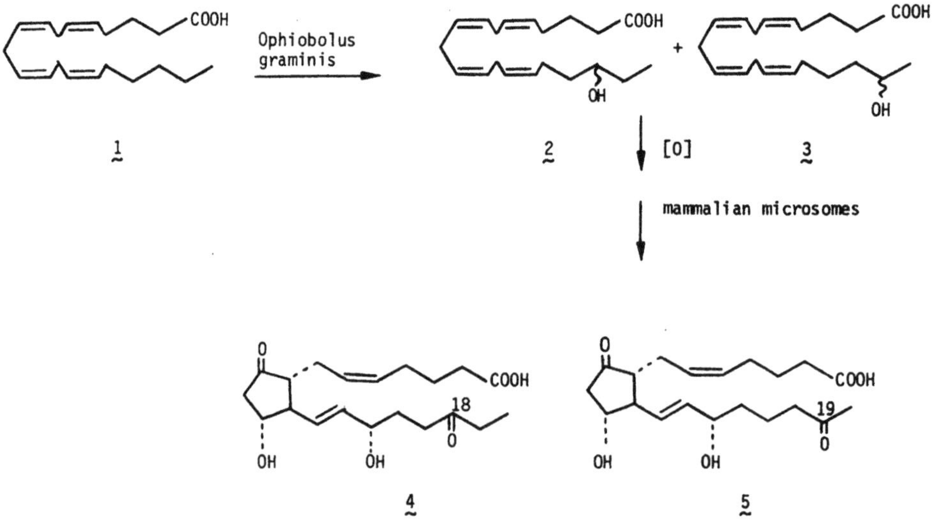

Fig. 4 Hydroxylation of arachidonic acid to 18ξ- and 19ξ-hydroxyarachidonic acid

hydroxyarachidonic acid in 17% yield (*3*). The biochemical conversions of these hydroxyarachidonic acids to the corresponding hydroxyprostaglandins were accomplished in low yields. Therefore, the individual hydroxyarachidonic acids were chemically oxidized with Jones' reagent to their respective ketones. Incubation of the oxoarachidonic acids with bull seminal vesicle microsomes cyclized them to 18-oxoPGE$_2$ (*4*) in 12% yield, and 19-oxoPGE$_2$ (*5*) in 11% yield.

2.2 Microbial Reactions in the Total Synthesis

A number of elegant total synthetic schemes have been decribed. These have been conveniently cataloged and reviewed[18,19,20]. The application of microbial reactions in any of these total synthetic schemes are usually at those steps where an asymmetric center is being introduced. PGF$_{2\alpha}$ has five asymmetric centers. Four of these asymmetric centers are common to all the primary prostaglandins. Of these, two asymmetric centers are associated with the attachment of the two aliphatic side chains to the cyclopentane ring, at C-8 and C-12. Two asymmetric centers are associated with the hydroxy groups at C-11 and C-15. The fifth asymmetric center at C-9 is involved in the interconversion of PGE to PGF.

There are numerous approaches to the total synthesis of prostaglandins. These have been catalogued by Bindra and Bindra[18] into three major classes. 1) Total sythesis where the cyclopentane ring is formed from an acyclic precursor. 2) Total synthesis in which the cyclopentane ring is performed and the two aliphatic side chains are attached stereospecifically. 3) Total synthesis using a bicyclic precursor which controls the stereochemistry of the α and ω side chains. Most of the chemical total syntheses not only gave dl-mixture as the final products, but also produced various other isomers en route, that had to be separated. To overcome some of these isomeric problems, several alternatives were used: the resolution of a racemic intermediate, preferably at an earlier stage in the total synthesis; the use of chiral reagents to induce stereoselective synthesis (this alternative requires the preparation of optically active derivatives of the chemical reagents); the use of an optically active starting material (this approach is limited to availability of optically active starting materials); and finally, the use of microbial reactions. Microbial reactions offer an alternative that appears promising, but has not been thoroughly studied.

Sih et al. are the major proponents of the use of microbial reactions in the total synthesis of prostaglandins. In an elegant series of papers, Sih progressively described the total synthesis of (+)-15-deoxyprostaglandin E$_1$[28], (—)-11-deoxyprostaglandin E$_1$[30], (—)-prostaglandin E$_1$ (PGE$_1$)[29,31], and (—)-prostaglandin E$_2$ (PGE$_2$)[32]. Through this series of papers, Sih gradually refined his "bioorganic total synthesis" of prostaglandins[33,34]. These bioorganic syntheses use both chemical and microbiological reactions.

The construction of the prostanoic acid nucleus is based on the conjugate addition of an eight carbon unit (octenyl synthon) (Fig. 5, *6*) as a vinyl copper reagent to the β-carbon of the conjugated enone system in 2-(6-carbomethoxyhexyl)-4-hydroxy-2-cyclopenten-1-one (*4*) (cyclopentyl synthon). The success of the "bioorganic total synthesis" is based on the stereospecific control of the chirality of the C-4 hydroxyl group (C-11, by prostaglandin numbering). The stereochemistry of the C-4

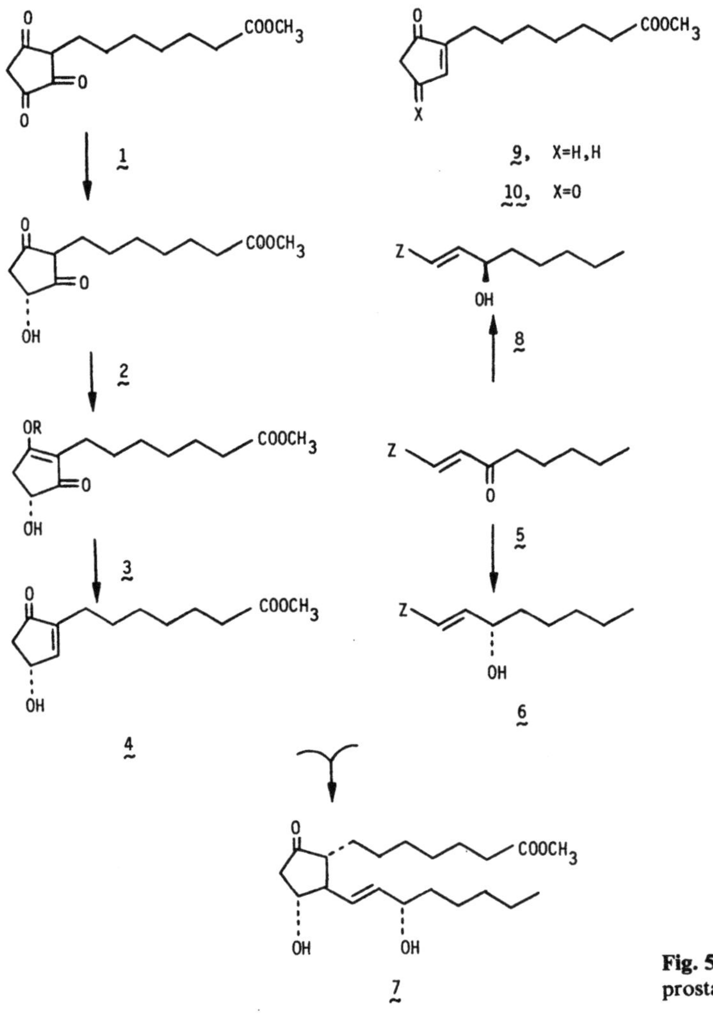

Fig. 5 Construction of the prostanoic acid nucleus

hydroxy group in the cyclopentyl synthon is critical, not only in that it will eventually become the C-11 hydroxy group in the final prostanoic acid, but that its stereochemistry dictates the eventual stereochemistry of the C-8 and C-12 side chains during the vinyl copper conjugate addition reaction. In the prostanoic acid molecule, the C-8, C-11, and C-12 positions are contiguous on the cyclopentane ring and exist in the more stable *trans-trans* arrangement. Thus, the control of the C-11 hydroxy stereochemistry eventually dictates the stereochemistry at C-8 and C-12.

Several alternatives were evaluated for the preparation of the octenyl synthon, 1-octen-3-ol (*6*)[32,34,35]. The reduction of the ketone *5* (Z = H) by various chemical methods yields racemic alcohols *6* (Z = H) which could be resolved via the (—)-α-methylbenzylamine salt of the corresponding hemiphthalate. Another alternative is the microbial reduction of 1-octen-3-one (*5*, Z = H). The resistence of an α,β-unsaturated ketone system to the microbial reduction was overcome by introduction an electronegative substituent on the α,β-unsaturated ketone system. Thus, 1-iodo-1-octen-3-one (*5*, Z = I) was stereospecifically reduced to (+)-3(S)-iodo-1-octen-3-ol (*6*, Z = I) by the washed cells of *Penicillium decumbens* or *P. vinaceum*, in both instances, in about 10% yields.

Under similar experimental conditions, *Aspergillus ustus* gave (—)-3(R)-iodo-1-octen-3-ol (*8*, Z = I) in 12% yield. Some 83 microorganisms of the classes *Ascomycetes*, *Phycomycetes* and *Fungi imperfecti* were found capable of stereospecific reduction of the ketone to either the R or S alcohol.

Several microbiological processes were evaluated for the preparation of the desired (+)-4(R)-hydroxy-2-cyclopenten-1-one (*4*) (cyclopentyl synthon). The stereospecific hydroxylation of 2-(6-carbomethoxyhexyl)-2-cyclopenten-1-one (*9*) was known to give material of low optical purity and produced degradative side products[34]. The microbial reduction of 2-(6-carbomethoxyhexyl)-2-cyclopentene-1,4-dione (*10*) with Baker's yeast reduced one of the two ketones and the double bond. However, the sterically more hindered C-1 ketone was reduced.

The problem was resolved by studying the microbial reduction of 2-(6-carbomethoxyhexyl)cyclopentane-1,3,4-trione (*1*) regioselectively at the C-4 ketone. *Dipodascus uninucleatus* regio- and stereospecifically reduced the trione (*1*) at the C-4 ketone to (+)-2-(6-carbomethoxyhexyl)-4(R)-hydroxycyclopentane-1,3-dione (*2*) in 75% yield[28, 34, 36]. The optically active dione was converted to the enol benzoate (*3*, R = —CO—Ph), or the isopropyl enol ether [*3*, R = $—CH(CH_3)_2$], and then reduced with sodium bis-(2-methoxyethoxy)aluminium hydride to yield the cyclopentyl synthon (*4*)[31, 34, 37, 38].

The conjugate addition of the ethoxyethyl ether *6* as the lithio cuprate to *4* afforded the natural (—)-PGE_1 methyl ester. The stereochemistry of the C-4 hydroxy group in the cyclopentyl synthon (*4*) dictated that the conjugate addition of the vinyl copper reagent (octenyl synthon) (*6*) at C-12 occurred on the β-side, resulting in a C-11, C-8, *trans-trans* configuration of the three contiguous chiral centers.

The microbial reduction of 2-(6-carbomethoxylhexyl)-cyclopentane-1,3,4-trione (*1*) with *Mucor rammanianus* afforded the (—)-4(S) alcohol, e.g. (—)-2-(6-carbomethoxyhexyl)-4(S)-hydroxycyclopentane-1,3-dione, in 43% yield[34, 36]. Some 67 microorganisms were found capable of regio- and stereospecific reduction of the trione (*1*). Development studies at the Miles Laboratories has improved the yields of scale-up processes[39]. By continuously controlling the carbohydrate concentration and the pH of the medium, along with periodic addition of substrate throughout the reaction, 35–80 g of substrate per 10 l were microbiologically reduced to the desired product in 72 h at about 70% yields.

For the preparation of PGE_2, the corresponding 2-(6-carbomethoxy-*cis*-2-hexenyl)-cyclopentane-1,3,4-trione was chemically synthesized[32, 34]. The rate of microbial reduction of the free acid was very slow. However, the methyl ester was readily reduced by *Dipodascus uninucleatus* in 24 h to (+)-2-(6-carbomethoxy-*cis*-2-hexenyl)-4(R)-hydroxycyclopentane-1,3-dione in 60% yield.

The hydrolysis of the conjugated addition products, PGE_1 methyl ester and PGE_2 methyl ester, to the free acids were achieved using *Rhizopus oryzae*. The free acids were obtained in 90% and 60% yields, respectively[31, 32, 34].

In total, three microbial reactions were used in Sih's "bioorganic total synthesis". Two of these microbial reactions were concerned with the introduction of chiral centers. Since one of these chiral centers dictated the chirality of two eventual chiral centers, all four of the chiral centers in the total synthesis of PGEs were microbiologically directed.

Weiss et al. at Lederle prepared a series of 15-deoxy-16-, 17-, and 20-hydroxy-prostaglandins via the 1,4 addition of lithium trialkyl *trans*-alkenyl alanate (Fig. 6, *6*) (octenyl synthon) to an appropriate cyclopentenone (*5*) (cyclopentyl synthon)[40]. The 15-deoxy-16ξ-hydroxy-PGE congeners exhibited good potencies when tested in various biological assays. For stereospecific synthesis, it was necessary to resolve the octenyl synthon. The racemic alcohol, 1-octyn-4-ol (*1*, R = H) was converted to the half-phthalates. Attempts to resolve the mixture using such bases as brucine, dehydroabietylamine or α-methylbenzylamine were ineffective. Using en-

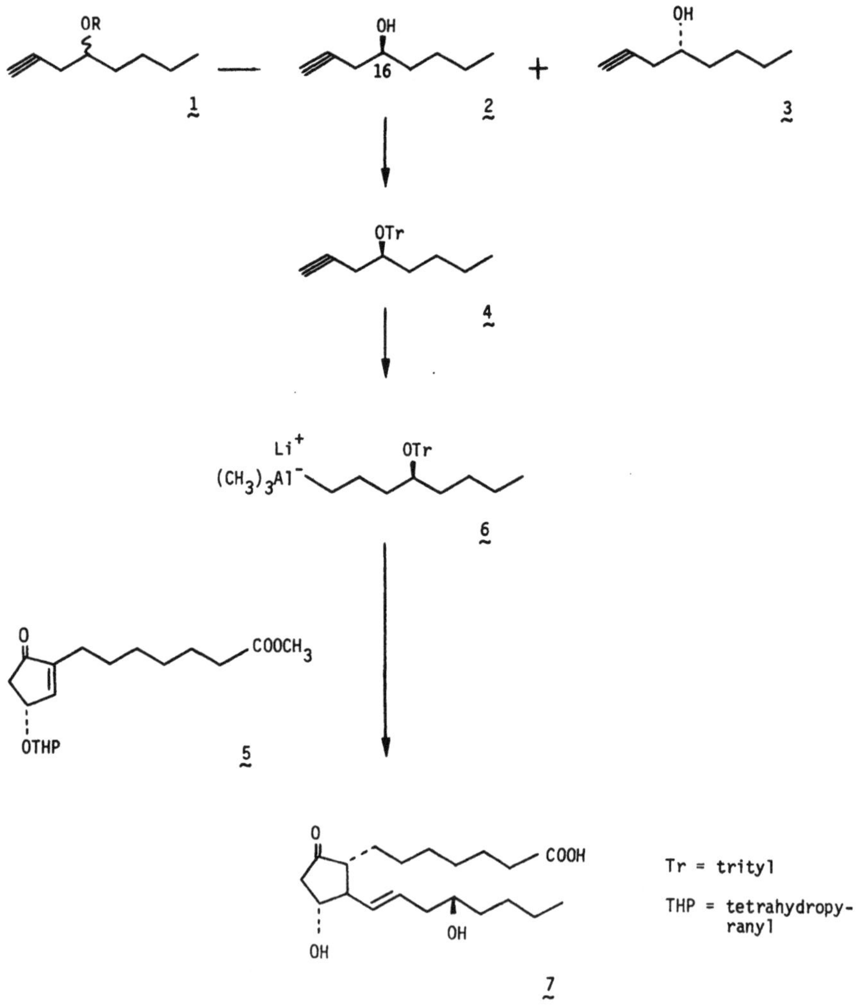

Fig. 6 Preparation of 15-deoxy-16, 17-, and 20-hydroxyprostaglandins

zymatic methods, the benzoate esters of the racemic 1-octyn-4-ol (*1*, R = —CO—Ph) was prepared. Screening identified *Rhizopus nigricans* R70 as being capable of selectively hydrolyzing the S benzoate ester to (—)-1-octyn-4(S)-ol. (*2*) in 14.5% yield[41].

The resolved (—)-1-octyn-4(S)-ol was converted to the crystalline half-phthalate-(—)-α-methylbenzylamine salt. This crystalline salt was used as seed crystals for the subsequent chemical resolution of (—)-1-octyn-4(S)-ol. From the filtrate was obtained the (+)-1-octyn-4(R)-ol (*3*).

A group at Teijin Limited prepared the same cyclopentyl synthon that Sih and many other groups used in their total synthesis of prostaglandins. The key intermediate, 2-(6-carboxyhexyl)-4(R)-hydroxycyclopenten-1-one (Fig. 7, *2*), was prepared by the direct microbial hydroxylation of 2-(6-carboxyhexyl)-2-cyclopenten-1-one (*1*)[42,43]. The allylic methylene position is activated to microbial hydroxylation as was demonstrated by the hydroxylation of cinerone (*3*) to cinerolone (*4*) by *Aspergillus niger*[44]. Several *Aspergillus* species (*A. niger* ATCC 9142 and IFO 6428, *A. tamarii* ATCC 1005, and *A. flavus* ATCC 12073) were found capable of microbiologically hydroxylating *1*. *Aspergillus niger* ATCC 9142 hydroxylated *1* to give *2* in 67% yield[42]. However, the (+)-2-(6-carboxyhexyl)-4(R)-hydroxy-2-cyclopenten-1-one was of low optical purity[34, Ref. 3]. Under preferred conditions, the yields ranged from 70–95%[43].

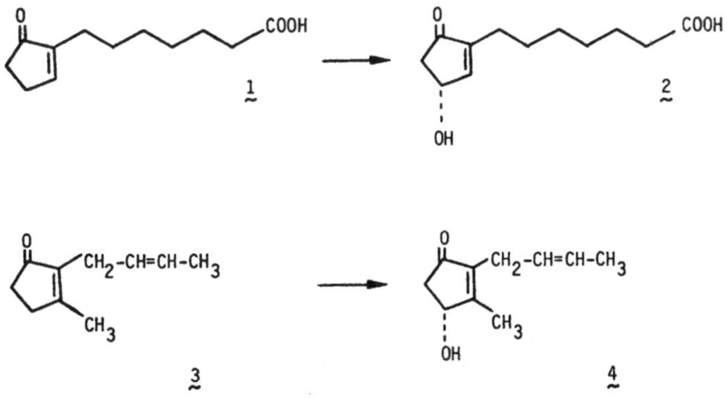

Fig. 7 Direct microbial hydroxylation of the Key intermediate

As substrate, the free acid was more readily hydroxylated than the methyl ester. Incubation times of greater than 40 h resulted in β-oxidation of the side chain to the bisnor-derivative. When the methyl ester was used as the substrate, prolonged incubation also hydrolyzed the ester group to the free acid.

Miyano *et al.* developed a total synthesis of prostaglandins involving the ring closure of the C-8, C-12 bond from an acyclic precursor (Fig. 8, *1*)[45,46,47]. This total synthetic approach involving relatively few steps and good yields is attractive, expecially for the manufacture of analogs. One of the analogs produced is (+)$\Delta^{8(12)}$-15-dehydroprostaglandin E_1 (*4*)[45,48]. Chemical reduction of the C-15 ketone group afforded a pair of *cis* and a pair of *trans* isomers with respect to the C-11 and C-15 hydroxyl groups. Microbiological screening was undertaken to look for microorganisms capable of regio- and stereoselective reduction of the C-15

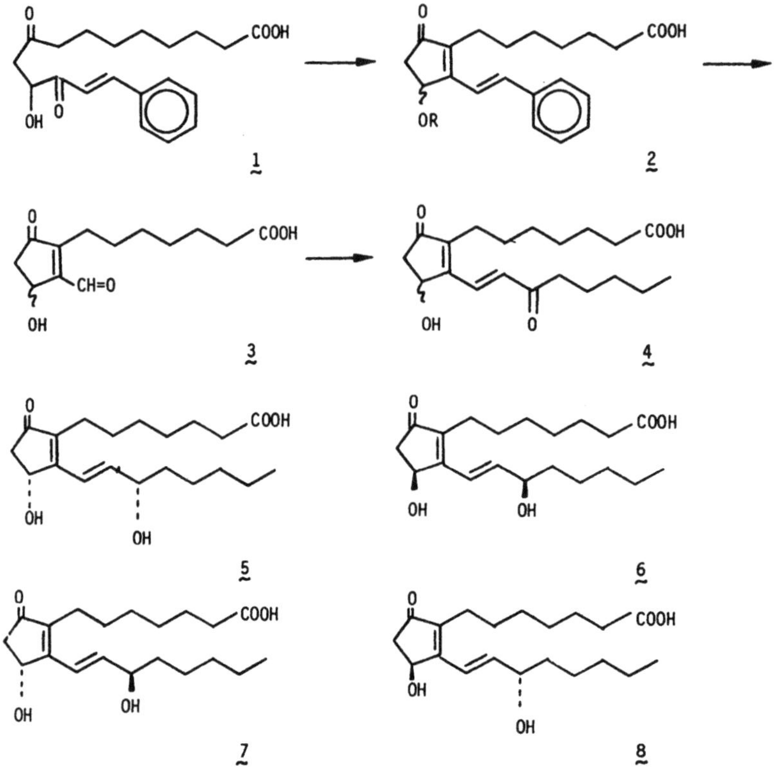

Fig. 8 Ring dosure of the C-8, C-12 bond from an acylic precursor

ketone[49,50]. Since the C-11 hydroxyl group is racemic, any reduction of the C-15 ketone could lead to 4 possible isomers (*5*, *6*, *7*, or *8*). To obtain the desired natural *cis* diol (*5*), two resolutions would be achieved in one step, the regio- and stereoselective reduction of the C-15 ketone of one of the C-11 racemates.

Flavobacterium sp. NRRL B-3874 reduced *4* to the *trans* diol *7* in 30% yield[49,50,51,52]. *Arthobacter* sp. NRRL B-3873 produced the same products. *Pseudomonas* sp. NRRL B-3875 reduced *4* to the *trans* diol *8* in 24% yield. *Rhodotorula glutinis* reduced *4* to a dl-*trans* diol.

Flavobacterium sp. NRRL B-5641 did reduce *4* to a *cis* diol. However, analysis of the product showed it to be an 1:1 mixture of the *cis* diols *5* and *6*. The reaction parameters for the *cis* reduction were more exacting than those used for the *trans* reductions.

The studies indicated a high degree of stereospecificity of the ketone hydrogenase activity in some bacteria. The presence of the C-8, 12 double bond was a necessary structural requirement for the ketone-hydrogenase activity. Reduction of the C-13, 14 double bond was a common side reaction.

Microbiological resolution of the intermediate 2-(R = H) was achieved by preparing the acetyl, propionyl or isobutyl racemic esters. *Saccharomyces* sp. 1375-143 stereoselectively hydrolyzed the (—)-R ester to yield the C-11(R) alcohol in 52% yield[53]. The optically active (—)-R alcohol *2* was used to prepare (—)-PGE_1 [42].

A group at Teijin studied the regio- and stereoselective hydrolysis of a 1:1 mixture of *cis* and *trans* 3,5-diacetoxycyclopent-1-ene (Fig. 9, *1*). Using baker's yeast, 3 products were isolated; the *trans* 3(R),5(R)-diacetoxycyclopent-1-ene (*2*), the *trans* 3(R)-acetoxy-5(R)-hydroxycyclopent-1-ene (*3*), and a mixture of predominantly *trans* 3(S),5(S)-dihydroxycyclopent-1-ene (*4*)[54]. This is probably the first evidence of an asymmetric half-hydrolysis of a simple molecule with two equivalent ester functions.

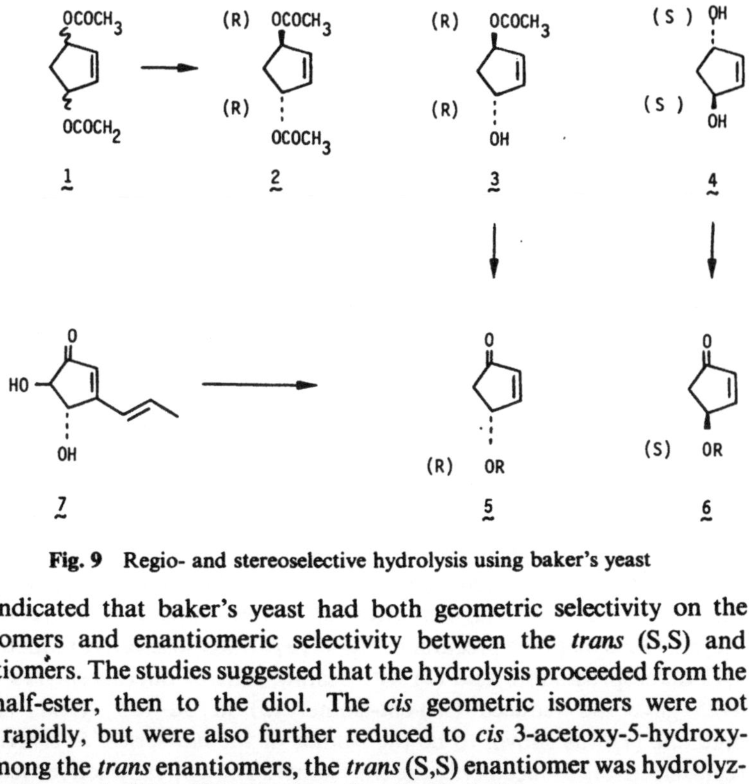

Fig. 9 Regio- and stereoselective hydrolysis using baker's yeast

Rate studies indicated that baker's yeast had both geometric selectivity on the *cis* and *trans* isomers and enantiomeric selectivity between the *trans* (S,S) and *trans* (R,R) enantiomers. The studies suggested that the hydrolysis proceeded from the diester, to the half-ester, then to the diol. The *cis* geometric isomers were not only hydrolyzed rapidly, but were also further reduced to *cis* 3-acetoxy-5-hydroxy-cyclopentane. Among the *trans* enantiomers, the *trans* (S,S) enantiomer was hydrolyzed more rapidly. By controlling the reaction (incubation) time, any of the three hydrolyzed products could be optimized. At 17 h incubation, the yield of 3(R)-acetoxy-5(R)-hydroxycyclopent-1-ene (*3*) was 11.5%.

The 3(R)-acetoxy-5(R)-hydroxycyclopent-1-ene (*3*) was chemically transformed to a protected 4(R)-hydroxycyclopent-2-en-1-one (*5*, R = a siloxy group)[55]. This optically active intermediate is equivalent to the prostaglandin synthon (*5*, R = cumyloxy group) Stork and Isobe used in their conjugate addition-enolate trapping approach for synthesis of natural prostaglandins[56]. The 3(S),5(S)-dihydroxy-cyclopent-1-ene (*4*) was chemically transformed to a protected 4(S)-hydroxycyclopent-2-en-1-one (*6*). This optically active synthon is useful for synthesis of 11-*epi*-prostaglandins.

Corey and Noyori[57], Fried and J. C. Sih[58], and Partridge et al.[59] all used a cyclopentane precursor in their total synthesis of prostaglandins. The prostaglandin synthrons *4* and *5* (Fig. 10) are common intermediates in the Corey and Partridge total synthesis, while the prostaglandin synthon *6* is a common intermediate in both the Corey and Fried total synthesis.

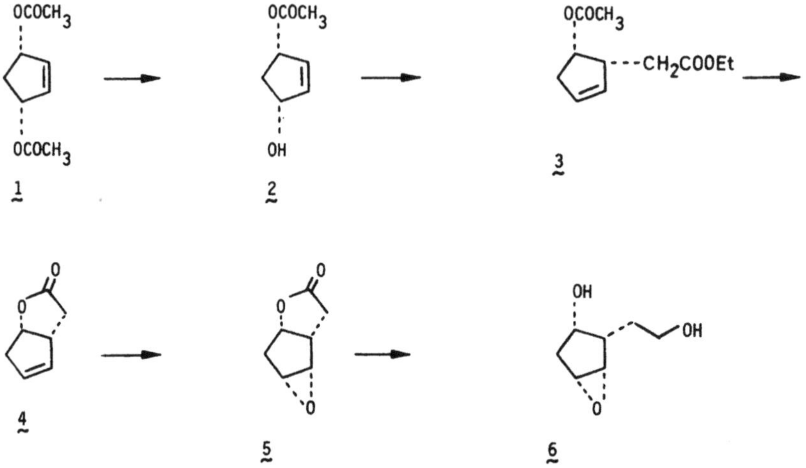

Fig. 10 Cyclopentane precursor in total synthesis of prostaglandins

Takano *et al.*[60] reported on an alternate asymmetric synthesis of the lactone *4*. Non-chiral *cis*-3,5-diacetoxycyclopent-1-ene was regioselectively hydrolyzed by *Bacillus subtilis* var. *niger* to *cis* 3-acetoxy-5-hydroxycyclopent-1-ene (*2*) in 56.1% yield. The chiral monoacetate (*2*) was chemically transformed to the lactone *4*. The lactone *4* possessed the desired absolute configuration with 35% optical purity.

A group at Ayerst synthesized a series of 11-deoxyprostaglandins by reductive dehalogenation with fragmentation of a C-C bond in the cyclobutane ring in a bicyclo [3.2.0] heptane precursor (Fig. 11, *1*)[61]. Regioselective reduction of the racemic 9,15-diketoprostanoic acid (*2*) with resting cells of *Saccharomyces cerevisiae* ATCC 4125 gave the reduced (+)-*ent*-isomer (*3*) in 30–40% yields[62]. The recovered diketone (*4*) was shown to be the isomer of the *nat*-series. *S. cerevisiae* ATCC 4125 regioselectively reduced only the *ent*-isomer. The optically active alcohol was oxidized to the *ent*-diketone. The *ent*-diketone and the recovered *nat*-diketone (*4*) were reacted individually with Grignard reagents to yield a series of isomeric 15-methyl tetrahydro PGA_1.

Using a modified approach of the Corey bicycloheptane total synthesis, a group at ICI has synthesized a series of 15-dehydroprostaglandins[63]. The C-15 ketone group was microbiologically reduced to the C-15(S) alcohol by various Basidiomycetes, especially *Trechispora brinkmanii* CMI 80,439[64].

The final step in most of the total synthetic schemes is the hydrolysis of the ester groups used to protect the carboxylic acid function during the entire total synthetic process.

Sih et al. used baker's yeast[28,29,30,33] and *Rhizopus oryzae*[31,32,34] in their series of "bioorganic" total synthesis of prostaglandins. Baker's yeast was also used to hydrolyze the ester groups in a series of Δ^2-prostaglandins[65].

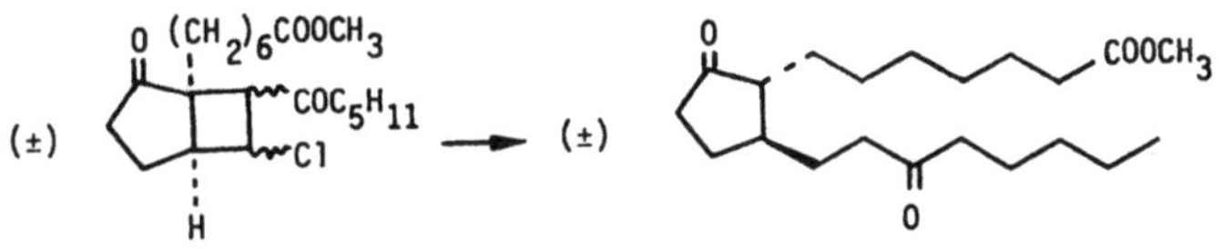

1 2

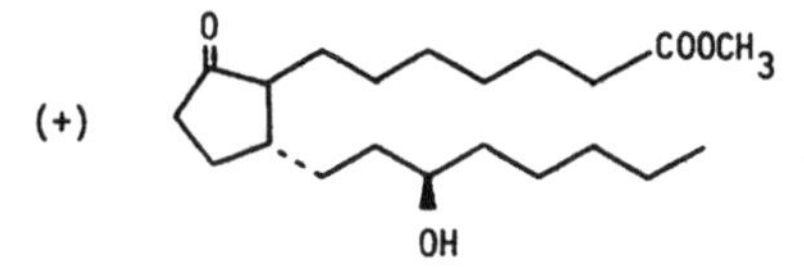

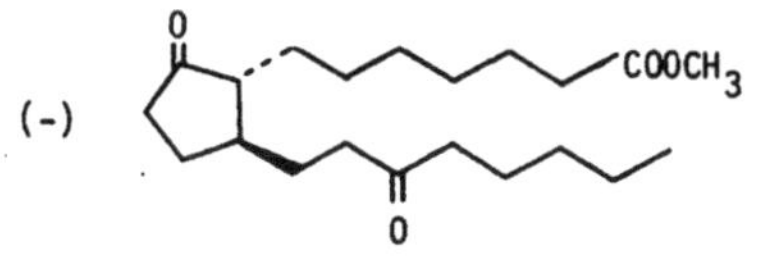

3 4

Fig. 11 Reductive dehalogenation with fragmentation of a C-C bond

The group at Upjohn found some 80 microorganisms capable of hydrolyzing the ester groups from the natural prostaglandins[66], and a series of 15-alkyl ether prostaglandins[67]. Microbial lipase has been used to hydrolyze the ester groups in prostaglandins[68]. The microbial lipase may also be fixed to a solid surface.

The gorgonian coral, *Plexura homomalla* is known to contain prostaglandins[69]. The esterase components were extracted from both the R forma and the S forma of *P. homomalla* and were used to hydrolyze prostaglandin esters[70]. The acetone-isoluble esterase can be recovered and reused.

The gorgonian coral, *P. homomalla*, forma R, contains 15-*epi*-PGA_2 and 15-*epi*-PGA_2 acetate methyl ester. The 15-*epi*-PGA_2 has been chemically converted to natural prostaglandins[71]. The diester, 15-*epi*-PGA_2 acetate methyl ester, is present in amounts greater than 600% those of 15-*epi*-PGA_2. In order to utilize this diester, it was hydrolyzed to the free alcohol acid in 63% yield using *Corynespora cassiicola* IMI 56007[72].

In a slightly different application of microbial synthesis, Mitscher and coworkers synthesized the same prostaglandin synthon (Fig. 9, *5*, R = acetyl) that Stork and Isobe[56] used for their total synthesis of prostaglandin. Miura and coworkers had prepared this same intermediate by asymmetrically selectively half-hydrolyzing a mixture of *cis* and *trans* 3,5-diacetoxycyclopent-1-ene (*1*). Mitscher et al. noted that terrein (*7*), a metabolite of *Aspergillus fischerii*, has the corrected absolute configuration at C-4, and can be used as a cyclopentyl synthon. Terrein was chemically converted to 4(R)-acetoxycyclopent-2-en-1-one (*5*) of high optical purity in 33% yield[73].

2.3 Microbial Transformations of Prostaglandins

The third sub-category are studies where microbial reactions are used to transform an intact prostaglandin molecule to some desired prostaglandin derivative. The natural prostaglandins are relatively unstable molecules and are labile to many chemical reactions. However, they are ideal substrates for the mild conditions of microbial reactions.

Schneider and Murray[74] studied the microbial reduction of *nat*-PGE_1 and *nat*-PGE_2 to *nat*-$PGF_{1\alpha}$ and *nat*-$PGF_{2\alpha}$, respectively using baker's yeast. The microbial reductions were stereospecific, since no appreciable amounts of the C-9β epimers were detected by thin layer chromatography. When the methyl esters were used as substrates, the methyl ester groups were first slowly hydrolyzed, then the C-9 ketone functions were reduced to the C-9(S) alcohols.

When recamic PGE_1 and PGE_2 were reduced, both enantiomers were reduced, producing *nat*-$PGF_{1\alpha}$, *nat*-$PGF_{2\alpha}$, *ent*-$PGF_{1\beta}$, and *ent*-$PGF_{2\beta}$. The reduction rates were quite slow, and many undesired side reactions were evident. Screening of other microorganisms and use of different experimental conditions did not improve the yields (about 10%), although *Torulopsis* yeast also did reduce the C-9 ketone function.

Sih et al.[33] also noted that the ester groups in racemic 15-deoxyprostaglandin methyl ester was hydrolyzed by baker's yeast in 13 h. If the substrate was incubated for 48 h or longer, the C-9 ketone functions were reduced to the respective *nat*-15-deoxypyrostaglandin $PGF_{1\alpha}$ and *ent*-15-deoxyprostaglandin $PGF_{1\beta}$. A group at Teijin also used baker's yeast to reduce the C-9 ketone in 7-oxoprostaglandin E_1 methyl ester[75]. Though the yield for the example given was about 58%, the stereochemistry of the resultant C-9 alcohol was not stated. The microbial reaction was regioselective in reducing only the C-9 ketone and not the C-7 ketone. However, the methyl ester group was hydrolyzed.

Jiu, Hsu and Mizuba studied the microbial transformations of prostaglandins A_2 (Fig. 12, *1*)[76,77,78]. *Cephalosporium* sp. NRRL 5499 transformed prostaglandin A_2 (*1*) to 15-hydroxy-9-ketoprosta-5,13-dienoic acid (*2*). The microbial reduction of the C-10, 11 double bond appeared to be a common microbial reduction. Many fungi and some bacteria were found capable of reducing the C-10, 11 double bond in the PGA system. *Dactylium dendroides* NRRL 2575 of 2-day growth produced two transformation products, *2* and 9,15-diketoprosta-5-enoic acid (*4*). The formation of *4* was postulated to go through the intermediate *3*. *Dactylium dendroides* NRRL 2575 of 7-day growth produced two transformation products, *4* and 9,15-diketoprosta-5,8(12)-dienoic acid (*7*). The formation of compound *7* was postulated to go through the intermediates *5* and *6*. *Cunninghamella blakesleeana* ATCC 9245 produced two transformation products, *2* and 15(S),18-dihydroxy-9-ketoprosta-5,13-dienoic acid (*8*)[79]. The hydroxylation was postulated to occur after the reduction of the C-10, 11 double bond; *1* to *2* to *8*.

Extending on their studies of the microbial hydrolysis of 15-*epi*-prostaglandin A_2 acetate methyl ester, Greenspan and Leeming noted that *Corynespora cassiicola* IMI 56007 not only hydrolyzed the ester groups, but in a modified medium and with the incubation time extended from 1 h to about 20 h, reduced the C-10, 11 double bond[80].

In a study comparing the microbial reduction of prostaglandin A_2 and 15-*epi*-prostaglandin A_2, Greenspan and Leeming noted that a series of microorganisms of the *Corynebacterium*, *Pseudomonas* and *Streptomyces* genera were capable of reducing the C-10, 11 double bond in prostaglandin A_2. However, only the species of the genera *Pseudomonas* and *Streptomyces* appeared capable of reducing the same C-10, 11 double bond in 15-*epi*-prostaglandin A_2 [81]. The selectivity of the *Corynebacterium* species in reducing only the C-10, 11 double bond in natural

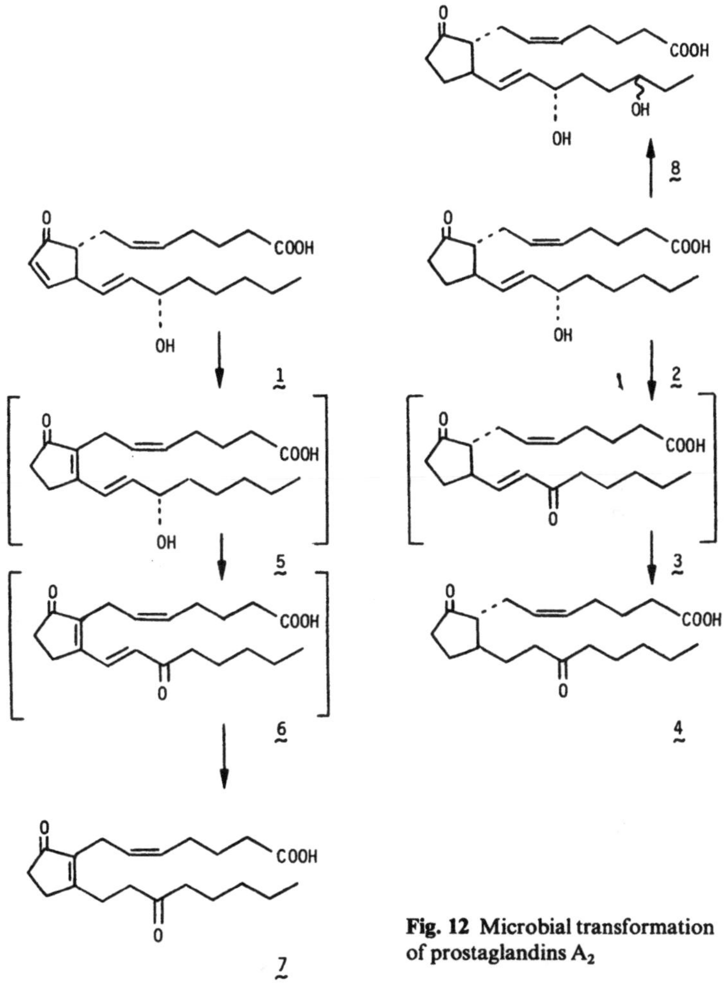

Fig. 12 Microbial transformation of prostaglandins A_2

prostaglandin A_2 apparently is dictated by the conformation of the C-15 hydroxyl group.

Hsu et al.[82] found four microorganisms capable of β-oxidizing natural and synthetic prostaglandins. *Penicillium* sp. M8904 transformed the postaglandin synthon (Fig. 13, *1*) to the dinor compound, racemic 5-(2-*trans*-styryl-2-hydroxy-5-oxo-cyclopentenyl)-*n*-pentanoic acid (*2*). *Mortierella* sp. M8872 stereospecifically β-oxidized *1* to the optically active dinor compound, (—)5-(2-*trans*-styryl-3(S)-hydroxy-5-oxo-cyclopentenyl)-*n*-pentanoic acid (*3*).

Penicillium sp. M8904 transformed prostaglandin B_2 (Fig. 14, *1*) to 2,3,4,5-tetranorprostaglandin B_2 (*2*), 18-hydroxyprostaglandin B_2 (*3*) and 19-hydroxyprostaglandin B_2 (*4*). Similarly, *Penicillium* sp. M8904 transformed prostaglandin A_2 (Fig. 15, *1*) to 15(S)-hydroxy-9-oxo-2,3,4,5-tetranorprosta-13-enoic acid (*2*), 9(R),-15(S)-dihydroxy-2,3,4,5-tetranorprosta-13-enoic acid (*3*), and 15(S)-hydroxy-9-oxo-prosta-5,13-dienoic acid (*4*).

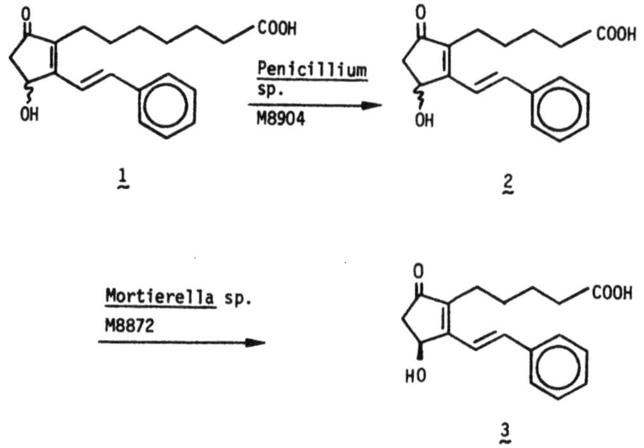

Fig. 13 *β*-oxidation of natural and synthetic prostaglandins

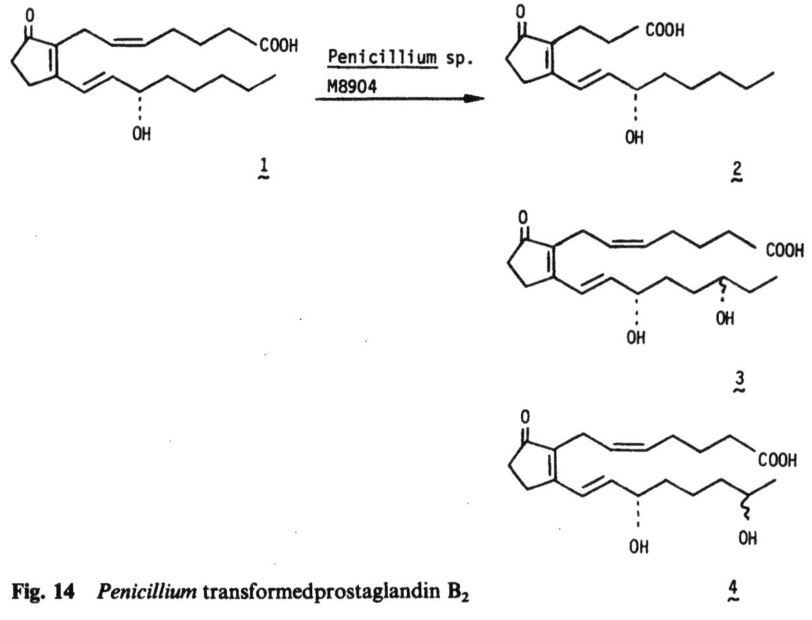

Fig. 14 *Penicillium* transformedprostaglandin B_2

Penicillium sp. M8904 apparently *β*-oxidized the natural prostaglandins to the tetranor derivatives, while the prostaglandin synthon was *β*-oxidized to the dinor derivative. The microbial *β*-oxidization occured only on the α-side chain.

Sebek *et al.*[83] described the hydroxylation of natural prostaglandin by *Streptomyces* UC5761. Prostaglandin $F_{2\alpha}$ was hydroxylated to the 18-hydroxyprostaglandin $F_{2\alpha}$ and the 19-hydroxyprostaglandin $F_{2\alpha}$ in 20–24% overall yields. Prostaglandin E_2 was hydroxylated to 18-hydroxyprostaglandin E_2, 19-hydroxyprostaglandin E_2, 18-hydroxyprostaglandin A_2, 19-hydroxyprostaglandin A_2, and 11-deoxy-19-hydroxyprostaglandin E_2. The formation of the hydroxylated prostaglandin A_2 probably involved a non-enzymatic dehydration of the substrate PGE_2 during the incubation and/or subsequent work-up. The formation of the 11-deoxy-19-hydroxyprostaglandin E_2 probably involved the subsequent microbial reduction of the C-10, 11 double bond of PGA_2 intermediate.

Additionally, the C-13, 14 double bond in 15-ketoprostaglandin $F_{2\alpha}$ was reduced by *Saccharomyces cerevisiae* in 46% yield.

Marsheck and Miyano[84] described the microbial hydroxylation of prostaglandin A_2 acetate methyl ester by members of the genus *Streptomyces*. The major microbial reactions observed were the hydrolysis of the two ester functions, reduction of the C-10, 11 double bond, and hydroxylation at the C-19 position of the prostaglandin molecule. The minor products were hydroxylated at the C-18 and C-17 positions. Also described was the ω-oxidation of the terminal C-20 methyl group to the dicarboxylic acid derivative.

Marx and Doodewaard[85] described the microbial hydroxylations of a series of synthetic prostaglandins by species of the *Streptomycetaceae* family. More than 150 synthetic prostaglandins were microbiologically hydroxylated at the C-18, C-19 and C-20 positions of the prostaglandin molecules. Many of the microbial transformation yields were in the 10–30% range. Some hydroxylations at the C-17 position were observed.

Lanzilotta et al.[86,87] described the microbial hydroxylation of both the *cis* and *trans* geometric isomers of dl-prosta-13-enoic acids using *Microascus trigonosporus*. Hydroxylation occurred at the C-18 position and the C-19 position in 20–35% yields. The hydroxylated *cis* products exhibited negligible optical properties. However, the hydroxylated *trans* products exhibited significant positive Cotton effects, indicating some degree of enantiomeric selectivity on hydroxylation with the *ent*-enantiomorph.

3 Classification by Microbial Reaction Types

In this section, the microbial reactions are reclassified according to reaction types. In this reclassification, the prostaglandin synthons with partial structures are catalogued according to the numbering of the carbon atoms in the final prostaglandin structure. The dehydration of the PGE-type of prostaglandins to the PGA-type of prostaglandins, and the isomerization of the PGA-type of prostaglandins to the PGB-type of prostaglandins are assumed to be artefacts of the experimental conditions used, and are not considered to be microbial reactions.

3.1 Microbial Cyclizations

Several articles described the direct microbial synthesis of prostaglandins or prostaglandin-like substances. Skarnes and Howard[21] prepared prostaglandins or prostaglandin-like substances from carbon and nitrogen sources using *Pseudomonas aeruginosa*. Beal et al. described the microbial synthesis of a series of prostaglandins and prostaglandin analogs from unsaturated fatty acids containing 1,4-diene systems[24]. These studies suggest the presence of a cylooxygenase-type of enzyme system in the members of the subphylum *Thallophyta*. This type of enzyme system has yet to be isolated from microbiological sources. Lipoxygenase-2 from soybean has been shown to catalyze the formation of prostaglandin from arachidonic acid[88].

3.2 Microbial Reductions

The types of microbial reductions observed are the microbial reduction of double bonds and the microbial reduction of ketones to alcohols. As yet, there is no evidence for the much sought for microbial reduction of the C-8, 12 double bond.

3.2.1 Microbial Reductions of Double Bonds

At present there is no direct evidence for the microbial reduction of the C-5, 6 double bond. Hsu et al.[82] have suggested that the C-5, 6 double bond is reduced by *Penicillium* M8904 in the process of the β-oxidation of natural prostaglandins to the tetranor derivatives.

The most common microbial reaction is the microbial reduction of the C-10, 11 double bond in the PGA system: Jiu et al.[76] have noted that this is a facile reduction for many microorganisms. Greenspan and Leeming did an extensive study on the microbial reduction of the C-10, 11 double bond. By modifying the medium and extending the incubation times, *Corynespora cassiicola* IMI 56007 not only hydrolyzed the ester groups[72], but also reduced the C-10, 11 double bond in PGA_2 acetate methyl ester[80]. Some selectivity was noted in that the *Corynebacterium* species only reduced the C-10, 11 double bond in natural PGA_2, but did not reduce the C-10, 11 double bond in 15-*epi*-PGA_2. Whereas, *Pseudomonas* and *Streptomyces* species reduced the C-10, 11 double bond in both PGA_2 and 15-*epi*-PGA_2 [81]. Sebek et al.[83] have also noted the microbial reduction of the C-10, 11 double bond by *Streptomyces* UC5761.

At present, there is no evidence for the microbial reduction of the C-13, 14 double bond in the presence of the C-15(S) hydroxyl group. All examples of microbial reduction of the C-13, 14 double bond are with compounds containing a C-15 ketone group (15-dehydroprostaglandins). The microbial formation of both 9,15-deketo-prosta-5-enoic acid (Fig. 12, *4*) and 9,15-diketoprosta-5,8(12)-dienoic acid (*7*) by *Dactylium dendroides* NRRL 2575 were postulated by Jiu et al.[76] to involve first the oxidation of the C-15 hydroxyl group to the ketone function, then reduction of the C-13, 14 double bond. Sebek *et al.*[83] described the reduction of the C-13, 14 double bond in 15-ketoprostaglandin $F_{2\alpha}$ by *Saccharomyces cerevisiae*. In their studies on the microbial reduction of the C-15 ketone in $(\pm)\Delta^{8(12)}$-15-dehydro-prostaglandin E_1 (Fig. 8, *4*), Marsheck and Miyano[50] noted that a wide variety of microorganisms also reduced the C-13, 14 double bond.

3.2.2 Microbial Reductions of Ketones

Schneider and Murray[74] described the reduction of the C-9 ketone in the microbial conversion of PGE to PGF using baker's yeast. The microbial reduction is stereoselective in that the microbial reduction occurred on the less hindered side of the C-9 ketone. The natural prostaglandins were reduced to the C-9(S) alcohols. Hsu et al.[82] noted that in the β-oxidation of PGA_2 by *Penicillium sp.* M8904 the C-9 ketone function was reduced to the C-9(R) alcohol (Fig. 15, *3*).

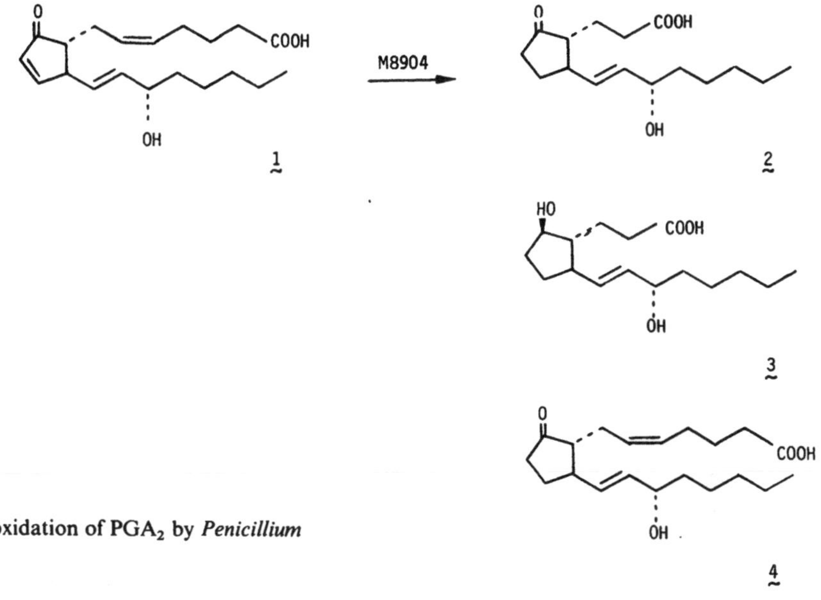

Fig. 15 *β*-oxidation of PGA_2 by *Penicillium*

Sih et al.[34] studying various alternatives for the preparation of the cyclopentyl synthon, noted that baker's yeast regioselectively reduced the more hindered C-1 ketone (C-9, by prostaglandin numbering) in 2-(6-carbomethoxyhexyl)-2-cyclopentene-1,4-dione. The stereochemistry of the hydroxyl group was not determined.

The one example describing the microbial reduction of the C-11 ketone used a cyclopentyl synthon as substrate. Sih et al.[28,34,36,38] noted that the microbial reduction of the C-4 ketone (C-11, by prostaglandin numbering) in 2-(6-carbomethoxyhexyl)-cyclopentane-1,3,4-trione (Fig. 5, *1*) by *Dipodascus uninucleatus* gave the C-11(S) alcohol (*2*). Using *Mucor rammanianus*, the cyclopentyl synthon was reduced to the C-11(R) alcohol.

A much sought after reaction is the stereospecific reduction of the C-15 ketone. Several groups described the stereospecific reduction of the C-15 ketone in prostaglandin analogs and the equivalent ketone position in an octenyl synthon.

Sih et al.[32,34] describe the microbial reduction of the octenyl synthon, 1-iodo-1-octen-3-one (Fig. 5, *5*, Z = I). *Penicillium decumbens* or *P. vinaceum* gave the (+)-3(S) alcohol (*6*, Z = I). This (+)-3(S) alcohol will eventually become the C-15(S) alcohol in the final prostaglandin molecule. Reduction of the octenyl synthon with *Aspergillus ustus* gave the (—)-3(R) alcohol (*8*, Z = I). The (—)-3(R) alcohol will eventually become the C-15(R) alcohol in 15-*epi*-prostaglandin.

Marsheck and Miyano[50] noted that the regioselective microbial reduction of the C-15 ketone in $(\pm)\Delta^{8(12)}$-15-dehydroprostaglandin E_1 (Fig. 8, *4*) gave principally the C-11, C-15 *trans* diols. *Flavobacterium sp.* NRRL B-3874 and *Arthrobacter sp.* NRRL B-3873 gave the *trans* diol of the 15-*epi*-prostaglandin series (*7*). *Pseudomonas sp.* NRRL B-3875 gave the *trans* diol of the 11-*epi*-prostaglandin series (*8*). Whereas, *Rhodotorula glutinis* gave a mixture of both *trans* diols. Under more exacting experimental conditions, *Flavobacterium sp.* NRRL B-5641 gave a 1:1 mixture of the *cis* diols; the *cis* diol of the natural series (*5*), and the 11,15-di*epi*-prostaglandin series (*6*).

Moore[64] has shown that the reduction of the C-15 ketone in a series of synthetic 15-dehydroprostaglandin E_1 using *Trechispora brinkmanii* CMI 80439 gave the C-15(S) alcohols.

3.3 Microbial Oxidations

The types of microbial oxidations described are the microbial oxidation of alcohols to ketones, the stereoselective microbial hydroxylation of the cyclopentane ring, the hydroxylation of the ω-side chain, the β-oxidation of the α-side chain, and the ω-oxidation of the terminal methyl group in the ω-side chain.

3.3.1 Microbial Oxidation of Alcohol to Ketone

Jiu, Hsu and Mizuba[76] documented the oxidation of the C-15 alcohols in PGA_2 and PGB_2 to the ketone functions by *Dactylium dendroides* NRRL 2575 (Fig. 12).

3.3.2 Microbial Hydroxylations

Several studies described the direct hydroxylation of the cyclopentane ring at the C-11 position. Based on an analogy to the hydroxylation of cinerone to cinerolone, Sih et al.[34], and Kurozumi et al.[42] described the direct hydroxylation of the cyclopentyl synthon, 2-(6-carboxy)-2-cyclopenten-1-one (Fig. 7, *1*) at the C-4 position (C-11 by prostaglandin numbering). Using various *Aspergillus* species, the hydroxylation was found to be partially stereospecific. Though the yields were good, the optical purities of the final products were low.

There were numerous examples describing the microbial hydroxylation of the intact prostaglandin molecule. The major positions of attack were the C-18 and C-19 positions. Sebek et al. described the hydroxylation of PGE_2 and $PGF_{2\alpha}$ by *Streptomyces sp.* UC 5761[83]. Jiu et al.[76] documented the hydroxylation of PGA_2 at C-18 by *Cunninghamella blakesleeana* ATCC 9245. While Marsheck and Miyano[84] described the hydroxylation of PGA_2 acetate methyl ester by species of the genus *Streptomyces*. Hsu et al.[82] described the hydroxylation of PGB_2 by *Penicillium sp.* M8904. Marx and Doodewaard[85] described the hydroxylation of prostaglandin analogs at the C-18, C-19 and C-20 positions by species of the Streptomycetaceae family. Lanzilotta et al.[86,87] described the hydroxylation of *cis* and *trans* (±)-prosta-13-enoic acid at C-18 and C-19 by *Microascus trigonosporus*. In most instances, minor hydroxylations were noted at the C-17 and C-20 positions.

Sih et al.[27] described the hydroxylation of arachidonic acid at C-18 and C-19 by *Ophiobolus graminis*. The oxygenated arachidonic acids were subsequently converted to oxygenated prostaglandins.

3.3.3 Degradative Microbial Oxidations

Hsu et al.[82] described the β-oxidation of PGA_2, PGB_2 and a prostaglandin synthon by *Penicillium sp.* M8904. The prostaglandin synthon was β-oxidized to a dinor derivative (Fig. 13), while the natural prostaglandins were β-oxidized to tetranor derivatives (Fig. 14 and 15). The β-oxidations occurred on the α-side chain.

Sih et al.[34], and Kurozumi et al.[42] studied the direct microbial hydroxylation of the cyclopentyl synthon, 2-(6-carboxyhexyl)2-cyclopenten-1-one, using *Aspergillus niger* ATCC 9142. The major side reaction noted was the β-oxidation of the carbomethoxyhexyl side chain. Marsheck and Miyano[84] documented the ω-oxidation of PGA_2 acetate methyl ester by various *Streptomyces* species.

3.4 Microbial Hydrolysis

Microbial reactions were used to hydrolyze the alkyl groups of carboxylic acid esters and to stereoselectively hydrolyze one of the ester groups in a racemic mixture of alcohol esters. In addition, microbial reactions were used to regioselectively hydrolyze one of two equivalent, or almost equivalent, ester functions on two separate chiral centers.

3.4.1 Optical Resolutions via Microbial Hydrolysis

Microbial hydrolysis of alcohol esters have been used to generate optically active centers in two different types of compounds with chiral carbons. One type of compounds is the usual mixture of racemic alcohol esters. In these compounds, the two alcohol esters are on the same chiral center. Stereoselective hydrolysis of one of the isomers would give an optically active free alcohol.

The other type of compounds contain two equivalent or almost equivalent alcohol ester functions on two separate chiral centers. Regioselective or regio- and stereoselective hydrolysis of one of the alcohol ester functions would give an optically active free alcohol.

Marsheck and Miyano[53] resolved (±)-7-(2-*trans*-styryl-3-hydroxy-5-oxocyclopentenyl)-*n*-heptanoic acid (Fig. 8, *2*). The esters of the prostaglandin synthon were resolved by stereoselective hydrolysis of the (—) R esters by *Saccharomyces sp.* 1375-143. The (—)-3-hydroxyl function is equivalent to the C-11(R) alcohol in the final prostaglandin molecule.

McGahren et al.[41] resolved the octenyl synthon, (±)-1-octyn-4-ol (Fig. 6, *1*) by stereoselective hydrolysis of the benzoate esters using *Rhizopus nigricans* R 70.

Takano et al.[60] noted that *cis* 3,5-diacetoxycyclopent-1-ene was regioselectively hydrolyzed by *Bacillus subtilis* var. *niger* to cis 3-acetoxy-5-hydroxycyclopent-1-ene (Fig. 10). The conformation of the C-3(S) acetoxy group eventually became the C-9(S) hydroxyl group in the final prostaglandin molecule.

Miura et al.[54] described the regio- and stereoselective microbial hydrolysis of a mixture of *cis* and *trans* 3,5-diacetoxycyclopent-1-ene by baker's yeast (Fig. 9). Among the products isolated were *trans* 3(R)-acetoxy-5(R)-hydroxycyclopent-1-ene (*3*) and a mixture of predominantly *trans* 3(S), 5(S)-dihydroxycyclopent-1-ene (*4*). In these products the conformations of the eventual C-11 hydroxyl groups were resolved or partially resolved. Both C-11 isomers were obtainable as the major product by controlling the reaction conditions.

Recently, Sih et al.[89] described the regioselective hydrolysis of 1α,4α-diacetoxy-2α,3β-dibenzoxymethylcyclopentane. *Aspergillus repens* gave exclusively the 1α-acetoxy,4α-hydroxy-2α,3β-dibenzoxymethylcyclopentane. The less hindered acetoxy group was hydrolyzed.

3.4.2 Microbial Hydrolysis of Esters

In most of the total synthetic schemes, alkyl ester groups were used to protect the carboxylic acid functions. Since prostaglandins are relatively labile compounds, mild conditions were needed to remove the protecting group. baker's yeast[28,29,30,33,65] and *Rhizopus oryzae*[31,32,34] were commonly used to hydrolyze the ester functions. Many other microorganisms were found capable of hydrolyzing the carboxylic acid ester functions[66,67,68], as well as the esterase components present in the gorgonian coral *Plexura homomalla*[70]. The acetate and methyl ester groups were hydrolyzed in PGA_2 acetate methyl ester using *Corynespora cassiicola* IMI 50076[72] and *Streptomyces* species[84].

4 Conclusion

Though the application of microbial reactions in prostaglandin chemistry is still in its infancy, a large number of microbial reactions have already been elucidated. Some of these microbial reactions have potential uses in large scale preparations of prostaglandin derivatives. For example, the scale-up experiments on the regio- and stereospecific reduction of 2-(6-carbomethoxylhexyl)cyclopentane-1,3,4-trione have been documented[39].

The appropriate place to use microbial reactions is dependent on the total synthetic scheme being used. Within a given total synthetic scheme, microbial reactions are generally useful in steps involved with the introduction of a chiral center. Except for the work of Sih et al., not many of the total synthetic schemes considered the use of microbial reactions in its original conception. Many of the applications of microbial reactions were based on an afterthought.

Another application of microbial reactions is for the preparation of derivatives of an intact prostaglandin molecule. Especially useful were the microbial hydrolysis of the ester groups of the carboxylic acid functions, and the hydroxylation of the prostaglandin molecule on the ω-side chain, particularly at C-18 and C-19 positions. Additionally, any total synthesis resulting in a 15-dehydroprostaglandin may be reduced microbiologically to the 15(S)-prostaglandin.

The feasibility of microbial reactions in prostaglandin chemistry has been documented. However, more research is needed to elucidate other microbial reactions. At present, there is no evidence for the microbial hydroxylation of the α-side chain, the dehydrogenation of an ethylene group to a double bond, the regioselective reduction of a C-11 ketone on an intact prostanoic acid to the C-11(S) alcohol, or the reduction of the C-8, 12 double bond. The latter would be a useful microbial reaction in that the synthesis of a $\Delta^{8(12)}$-prostanoic acid is much more readily achieved, since it has two less asymmetric centers.

In addition, much development research is needed to improve on the yields and optical purities of the presently known microbial reactions. These refinements and improvements would probably not be studied until a racemic prostanoic acid derivative has been demonstrated to have commercial potential. Then, the expenditures for development research can be justified. Also, at that point, it may be necessary to prepare the individual racemic components in high optical purity, because there are increasing evidences that each of the racemic compounds possesses completely different biological properties.

5 Acknowledgement

The author is indebted to C. H. Fay and M. S. Russell of the Corporate Information Department for collation of the references, and to G. Patterson for preparation of the manuscript.

6 References

1. Kurzrok, R., Lieb, C. C.: Proc. Soc. Exp. Biol. Med. *28*, 268 (1930)
2. Goldblatt, M. W.: J. Soc. Chem. Ind., Lond. *52*, 1056 (1933)
3. Von Euler, U. S.: Arch. Exp. Path. Pharmacol. *175*, 78 (1934)
4. Von Euler, U.S.: J. Physiol., Lond. *88*, 213 (1936)
5. Von Euler, U. S.: Skand. Arch. Physiol. *81*, 65 (1939)
6. Bergström, S., Sjövall, J.: Acta Chem. Scand. *11*, 1086 (1957)
7. Bergström, S., Sjövall, J.: Acta Chem. Scand. *14*, 1693 (1960)
8. Bergström, S., Sjövall, J.: Acta Chem. Scand. *14*, 1701 (1960)
9. Bergström, S. et al.: J. Biol. Chem. *238*, 3555 (1963)
10. Bergström, S. et al.: Ark. Kemi *19*, 563 (1962)
11. Samuelsson, B.: J. Amer. Chem. Soc. *85*, 1878 (1963)
12. Bergström, S., Carlson, L. A., Weeks, J. R.: Pharmacol. Rev. *20*, 1 (1968)
13. Hinman, J. W.: Ann. Rev. Biochem. *41*, 161 (1972)
14. Samuelsson, B. et al.: Ann. Rev. Biochem. *44*, 669 (1975)
15. Weeks, R. J.: Ann. Rev. Pharmacol. *12*, 317 (1972)
16. Bergström, S., Danielson, H., Samuelsson, B.: Biochem. Biophys. Acta *90*, 207 (1964)
17. Van Dorp, D. A. et al.: Biochim. Biophys. Acta *90*, 204 (1964)
18. Bindra, J. S., Bindra, R.: Prostaglandin Synthesis, New York: Academic Press 1977
19. Szàntay, C. S., Novàk, L.: Synthesis of Prostaglandins, Budapest: Akademiai Kiado 1978
20. Mitra, A.: The Synthesis of Prostaglandin. New York: Interscience 1977
21. Skarnes, R. C., Howard, L. A.: U.S. 3,843,467 (Oct. 22, 1974)
22. Bobylev, R. V. et al.: U.S.S.R. 457, 475 (Jan. 25, 1975); Chem. Abstr. *83*, 26321u (1975)
23. Gandel, V. G. et al.: U.S.S.R. 461,731 (Febr. 28, 1975); Chem. Abstr. *83*, 41521c (1975)
24. Beal, III, P. F., Fonken, G. S., Pike, J. E.: U.S. 3,290,226 (Dec. 6, 1966); Belgian 659,983 (Aug. 19, 1965); Netherlands 6,502,097 (Aug. 20, 1965)
25. Watanabe, A., Ono, Y.: Japan, 72 37,030 (Sept. 18, 1972)
26. Iizuka, H., Ootumo, T., Youshida, K.: Japan. Kokai 77 64,483 (Nov. 22, 1975); Japan. Kokai 77 64,484 (Nov. 22, 1975)
27. Sih, C. J. et al.: J. Amer. Chem. Soc. *91*, 3685 (1969)
28. Sih, C. J. et al.: Chem. Commun. 240 (1972)
29. Sih, C. J. et al.: J. Amer. Chem. Soc., *94*, 3643 (1972)
30. Sih, C. J. et al.: Tetrahedron Lett. 2435 (1972)
31. Sih, C. J. et al.: J. Amer. Chem. Soc., *95*, 1676 (1973)
32. Heather, J. B. et al.: Tetrahedron Lett. 2313 (1973)
33. Sih, C. J. et al.: J. Amer. Chem. Soc. *97*, 857 (1975)
34. Sih, C. J. et al.: J. Amer. Chem. Soc. *97*, 865 (1975)
35. Sih, C. J.: U.S. 3,868,306 (Febr. 25, 1975)
36. Sih, C. J.: U.S. 3,773,622 (Nov. 20, 1973)
37. Sih, C. J., Heather, J. B.: U.S. 3,968,141 (July 6, 1976)
38. Sih, C. J., Heather, J. B.: U.S. 4,057,851 (Nov. 8, 1977)
39. Chang, L. T., Terry, C. A.: U.S. 3,925,156 (Dec. 9, 1975)
40. Floyd, M. G., Schaub, R. E., Weiss, M. J.: Prostaglandins *10*, 289 (1975)

41. McGahren, W. J. et al.: J. Org. Chem. *42*, 1659 (1977)
42. Kurozumi, S., Toru, T., Ishimoto, S.: Tetrahedron Lett. 4959 (1973)
43. Kurozumi, S., Toru, T., Ishimoto, S.: U.S. 3,892,630 (July 1, 1975); Netherlands 7,407,125 (Dec. 2, 1974); Japan 77 001,995 (Jan. 19, 1977)
44. Tabenkin, B. et al.: Appl. Microbiol. *17*, 714 (1969)
45. Miyano, M., Dorn, C. R.: Tetrahedron Lett. 1615 (1969)
46. Miyano, M., Dorn, C. R., Mueller, R. A.: J. Org. Chem. *37*, 1810 (1970)
47. Miyano, M., Stealey, M. A.: Chem. Commun. 180 (1973); J. Org. Chem. *40*, 1748 (1975)
48. Miyano, M., Dorn, C. R.: J. Org. Chem. *37*, 1818 (1972)
49. Miyano, M. et al.: Chem. Commun. 425 (1971)
50. Marsheck, W. J., Miyano, M.: Biochim. Biophysica Acta *348*, 263 (1974)
51. Colton, F. B., Marsheck, W. J., Miyano, M.: U.S. 3,687,811 (Aug. 29, 1972)
52. Marsheck, W. J., Miyano, M.: U.S. 3,799,841 (March 26, 1974)
53. Marsheck, W. J., Miyano, M.: Biochim. Biophysica Acta *316*, 363 (1973)
54. Miura, S. et al.: Tetrahedron *32*, 1893 (1976)
55. Tanaka, T. et al.: Tetrahedron *32*, 1713 (1976)
56. Stork, G., Isobe, M.: J. Amer. Chem. Soc. *97*, 6260 (1975)
57. Corey, E. J., Noyori, R.: Tetrahedron Lett. 311 (1970)
58. Fried, J., Sih, J. C.: Tetrahedron Lett. 3899 (1973)
59. Partridge, J. J., Chadha, N. K., Uskokovic, M. R.: J. Amer. Chem. Soc. *95*, 7171 (1973)
60. Takano, S., Tanigawa, K., Ogasawara, K.: Chem. Commun. 189 (1976)
61. Bagli, J. F., Bogri, T.: J. Org. Chem. *37*, 2132 (1972)
62. Bagli, J. F., Sehgal, S. N.: Tetrahedron Lett. 3329 (1973)
63. Brown, E. D. et al.: Chem. Commun. 642 (1974)
64. Moore, R. H.: Ger. Offen. 2,401,761 (July 18, 1974)
65. Hayashi, M. et al.: Ger. Offen. 2,460,285 (July 3, 1975)
66. Aken, U.F., Murray, H. C.: Ger. Offen. 1,937,678 (Jan. 29, 1970)
67. Lincoln, F. H., Pike, J. E., Youngdale, G. A.: Ger. Offen. 2,221,443 (Nov. 16, 1972)
68. Stadler, I. et al.: German 2,247,792 (April 12, 1973)
69. Weinheimer, A. J., Spraggins, R. L.: Tetrahedron Lett. 5183 (1969)
70. Daniels, E. G.: U.S. 3,840, 434 (Oct. 8, 1974)
71. Bundy, G. et al.: Ann. N.Y. Acad. Sci. *180*, 76 (1971)
72. Leeming, M. R. G., Greenspan, G.: U.S. 3,726,765 (April 10, 1973)
73. Mitscher, L. A., Clark, III, G. N., Hudson, P. B.: Tetrahedron Lett. 2553 (1978)
74. Schneider, W. P., Murray, H. C.: J. Org. Chem. *38*, 397 (1973)
75. Teijin, K. K.: Japan 77 128,354 (Oct. 27, 1977)
76. Jiu, J., Hsu, C. F. J., Mizuba, S.: Dev. Ind. Microbiol. *15*, 345 (1974)
77. Hsu, C. F., Jiu, J., Mizuba, S. S.: U.S. 3,788,947 (Jan. 29, 1974)
78. Hsu, C. F., Jiu, J., Mizuba, S. S.: U.S. 3,868,412 (Febr. 25, 1975)
79. Hsu, C. F., Jiu, J., Mizuba, S. S.: U.S. 3,856,852 (Dec. 24, 1974)
80. Greenspan, G., Leeming, M. R. G.: U.S. 3,928,134 (Dec. 23, 1975)
81. Greenspan, G., Leeming, M. R. G.: U.S. 3,930,952 (Jan. 6, 1976)
82. Hsu, C. F. J., Jiu, J., Mizuba, S.: Dev. Ind. Microbiol. *18*, 487 (1977)
83. Sebek, O. K., Lincoln, F. H., Schneider, W. P.: Abstracts, 5. Internat. Ferment. Symp., 17. 05. Berlin, West Germany 1976
84. Marsheck, W., Miyano, M.: U.S. 3,878,046 (April 15, 1975)
85. Marx, A. F., Doodewaard, S.: U.S. 4,054,595 (Oct. 18, 1977)
86. Lanzilotta, R. P. et al.: Appl. Environ, Microbiol. *32*, 726 (1976)
87. Lanzilotta, R. P.: U.S. 4,036,876 (July 19, 1977)
88. Bild, G. S. et al.: J. Biol. Chem. *253*, 21 (1978)
89. Sih, C. J., Huand, F. C.: J. Amer. Chem. Soc. *100*, 643 (1978)

Methanol as Carbon Source for Biomass Production in a Loop Reactor

U. Faust, W. Sittig
Hoechst A.G., 6000 Frankfurt/M. 80, Federal Republic of Germany

1 Introduction ... 64
1.1 Methanol-Utilizing Microorganisms ... 65
1.2 Requirements of Methanol Utilizing Strains ... 66
1.2.1 Oxygen Tension ... 67
1.2.2 Nutrient Medium ... 67
1.2.3 Methanol ... 67
1.2.4 Sterility ... 68
1.2.5 Carbon Dioxide ... 68
1.2.6 Cell Density ... 68
1.2.7 Temperature ... 68
2 Biological Requirements for the Reactor Design ... 68
2.1 Specification of the SCP Reactor ... 70
2.2 Periphery of a Bioreactor ... 71
2.3 Types Proposed ... 71
2.4 Bubble-Column Reactors ... 73
2.5 Loop Reactors ... 74
2.6 Breaking of Foam ... 79
2.7 Scale-Up of an Internal Loop Reactor ... 79
3 Consideration of Mass Transfer ... 80
3.1 Oxygen-Transfer Resistance ... 80
3.2 Gas-Liquid Interface ... 82
3.3 Oxygen Balance in a Methanol Process ... 84
3.4 Significance of Convection for Mass Transfer ... 86
3.5 Oxygen Transfer in the Downcomer ... 88
3.6 Gas- and Liquid-Mixing ... 89
4 Air-Lift Drive ... 90
4.1 Gas Content in the Riser ... 91
4.2 Pressure Losses ... 92
5 Discussion ... 94
7 Symbols ... 95
8 References ... 97

Literature concerning mass transfer in bubble columns and loop reactors contains numerous data on small-scale units. On the other hand, data on the production scale, i.e. reactor diameters exceeding 1 m, are scarce. The present work tries to apply the data that are available to the calculation of a loop reactor. The metabolic reaction itself is performed in each single cell. A careful consideration of the statistics in cell-medium systems is used as a necessary background for an economic reactor design. Based on this fact, the actual supply of each cell is discussed for the growth of *Methylomonas clara* on methanol.

Most of the literature takes for granted that the calculation of gas-liquid oxygen transfer, i.e. the transportation from the bubble to the liquid layer, describes the real problem. It is shown by reasonable assumptions that the further oxygen transport from the gas-liquid film to each individual microorganism must be considered as well. In addition, the importance of macro- and micro-mixing for the scale-up procedure are pointed out.

For a medium-sized reactor of 300 m^3 total volume, the main features of scale-up calculations are considered. Little data being published, the zones of critically low supply are studied using rough estimates. The advantages and limitations of an air-lift loop reactor compared to those of stirred tank reactors are discussed.

1 Introduction

The history of bioreactors has occurred in four stages. The first types were simple anaerobic vats for alcohol and yeast production with few requirements for oxygen and mixing. In the second stage, surface type reactors were developed for the production of primary metabolites such as acetic and citric acid. The third stage made use of stirred tanks for the aerobic submerged antibiotic production. Here non-Newtonian media required aseptic running and high energy input. SCP processes give rise to a fourth generation of reactors. These are characterized by large reactor volumes, the necessity of continuous running and a high productivity at low energy input.

The development of SCP processes demands extensive investigations of reactor design and operating conditions which lead to various types of new reactors. These will be dealt with in this chapter with special regard to methanol as a substrate.

The microbial conversion of methanol to biomass has been described by a number of authors[62]. The value of methanol as a substrate compared with that of hydrocarbons and carbohydrates is particularly suitable for investigations concerning reactor design for the following reasons:

The synthetic media used in processes involving methanol are chemically and physically well defined and give a completely mixed liquid;

the microorganism strains employed require a high oxygen and substrate supply;

the biomass from methanol is the most promising and economic way of large-scale SCP production in the near future. It will most likely be carried out in a working volume of vessels with several thousand cubic meters. For this task, scale-up considerations are necessary.

This chapter concentrates on the practical application of loop reactors to the protein production with methanol as a substrate.

The theoretical background of the loop reactor principle has been described by Blenke et al.[11,12].

1.1 Methanol-Utilizing Microorganisms

Although methanol, as well as other C_1-compounds, is no common biological substrate, a number of specialized microbial strains, including obligate and facultative methylotrophic bacteria, yeasts and fungi, have been isolated, the metabolism of which was described by Cooney[23], and Wagner[108] and comprises a) the intracellular oxidation of methanol via formaldehyde to CO_2 to generate energy and reduction equivalents and b) the biosynthesis of cellular material from formaldehyde along the serine or ribulosephosphate pathways:

$$CH_3OH \rightarrow HCHO \rightarrow HCOOH \rightarrow CO_2$$

$$\downarrow$$

C_3-compounds (e.g., glyceraldehyde-3-phosphate)

$$\downarrow$$

serine pathway or ribulose pathway

An overall equation of methanol bioconversion may be formulated as follows[2]:

$$a\,CH_3OH + b\,O_2 + c\,NH_3 + \text{minerals}$$
$$\rightarrow \times \text{ biomass} + y\,CO_2 + z\,H_2O + \text{by-products} + \text{heat.}$$

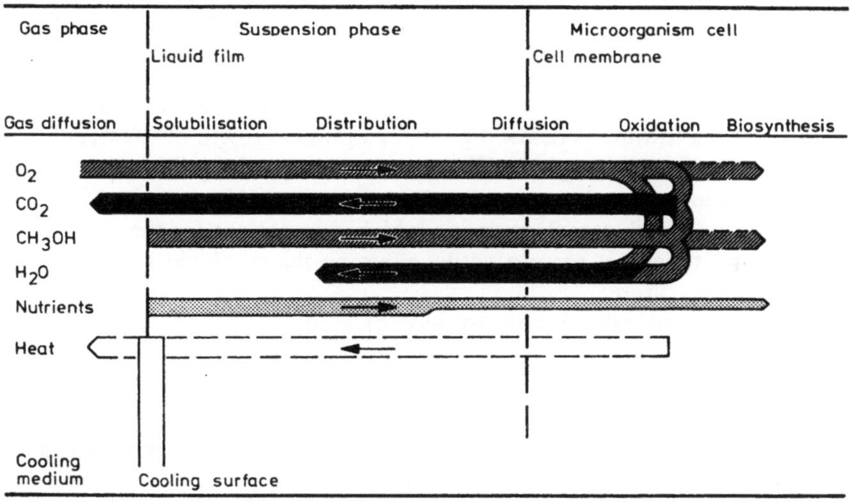

Fig. 1 Mass-heat transport in methanol processes for SCP

It is possible to carry out this reaction in fully synthetic. Newtonian aqueous media, which contain the necessary phosphate, sulfate, ammonium or nitrate, magnesium, potassium, calcium, ferrous or ferric salts, and trace elements. Other microorganisms require in addition natural growth factors such as biotin. Thermophilic strains are described by Harrison et al.[43], Cooney et al.[21-23] and Snedecor and Cooney[102]. In general yeasts and fungi give lower yields ranging from $Y_{x/s} = 0.2$ to 0.4[2,22,26] whereas bacteria reach Y_s-values of 0.4–0.5[39]. Theoretical yield factors[43] range from 0.83 to 0.46 depending on the biochemical pathway of

C_1-fixation[19]. SCP-production from yeast is less economic than from bacteria. The inferior productivities, lower yields, protein contents, and cultivating temperatures, of yeasts which are biologically determined, dominate production costs of protein to such a large extent[109] that only for special applications (e.g. acceptability, vitamin premium, registration problems) yeast from methanol can compete with bacteria. For bacterial SCP, technical problems of the reactor lay-out, however, prove to be more sophisticated.

1.2 Requirements of Methanol Utilizing Strains

All reported strains may be grown in one-stage continuous cultures. On an industrial scale, yeasts are often cultivated at dilution rates of around $0.1\ h^{-1}$ and at cell densities of 10 to $30\ g\ l^{-1}$ [61,12] whereas bacterial strains are cultivated at dilution rates of $0.2–0.5\ h^{-1}$ and cell densities of 10 to $50\ g\ l^{-1}$ [11,13]. Most chemostat laboratory studies are carried out, however, at much lower cell densities, i.e. $1–6\ g\ l^{-1}$. This fact restricts the value of those studies for scale-up considerations to small-cell densities. Yeast tolerate acidic conditions (i.e. pH values of 3.5 to 5)[12] whereas bacteria require the neutral pH range of 6.8-7.5[14,15].

Various authors report special characteristics of methylotrophic strains: Dostalek[27] observed extracellular protein production at low growth rates and methanol concentrations below $4\ g\ l^{-1}$ for optimal growth of *Methylomonas methanica*. Chalfan[18] and Haggström[41] observed extracellular polysaccharide production at high temperature, low growth rate and oxygen limitation. The presence of polysaccharide facilitates flocculation which is important for the ease of harvesting. At the same time, this has a negative effect on the rheological properties and mass transfer in the culture.

A stable mixed culture with no need of aseptic operation was found by Harrison[43]. Brooks[15] observed a reduction in the methanol yield when the substrate was added to a C-limited culture of *Pseudomonas* at intervals greater than 20 s. After each addition, an initial burst of growth was followed by a period of starvation accompanied by an increased CO_2 output. On the other hand, methanol at higher concentrations acted as a growth inhibitor.

Babij[7] observed CO_2 assimilation accompanied by stimulated growth at low methanol concentrations for *Pseudomonas*. Harrison[42] found methanol inhibition at concentrations exceeding 1%. This was also reported by Astuana[6] and Ogata[79]. MacLennan[71] found a significant yield reduction at oxygen pressures above 100 mm Hg. Gow[40] discovered a killing effect of *Pseudomonas* during external cooling, possibly due to thermal shock, to a rise in the CO_2 concentration or to a sudden pressure change. For a two-phase injector, Sittig[101] reported a negative influence of locally high shear forces upon the substrate yield of *Methylomonas clara*, resulting in an enhanced CO_2 production and heat evolution, the growth rate being the same.

Although in each case, these findings have been evaluated with individual strains, some common features of methanol-utilising microorganism cultures may be deduced from them which are important for the reactor lay-out.

In contrast to reaction rates of one-step chemical reactions, the microbial conversion kinetics of methanol do not exhibit substrate concentration proportionality. Biomass production is generally carried out under chemostat conditions at substrate concentrations which lie at the upper end of the Michaelis-Menten proportionality in order to achieve a maximum productivity. Thus, the overall reaction rate becomes independent of the substrate concentration, as long as a minimum threshold value is maintained throughout the culture. As can be deduced from the values given above, concentrations should be kept within the limits of a defined concentration range. Concentrations outside this tolerance, in many cases, lead to growth inhibition.

1.2.1 Oxygen Tension

Although a safe threshold concentration recommended for a sufficient oxygen supply is usually fixed at around 10% of atmospheric O_2 pressure[26,28], detailed measurements in the vicinity of respiring cells with microelectrodes[27] or by means of studies with isotopes[29] have not yet been reported. Kuraishi postulated a minimum oxygen concentration of 2 ppm[60]. In the case of methanol, oxygen limitations lead to undesirable reactions of the culture. According to Tsao, experimental microscale data indicate the oxygen supply in most cases to be insufficient[108]. An upper concentration limit of dissolved oxygen has been predicted but yet been probed only for a few strains[23]. Miura found that for all aerobic cultures, the respiration rate is independent of the O_2 concentration down to a limiting DOT value which is called the critical O_2 concentration.

1.2.2 Nutrient Medium

Optimum pH values for microbial growth were determined experimentally by means of the newly established, improved wash-out method in chemostat cultures as reported by Cooney[20]. These values have a reasonable tolerance range of ± 0.1 pH (in the case of most yeast strains, the operating pH range is much broader reaching more than 7.0 to 3.5).

As Pirt[83] has shown for batch cultures, a wide range of nutrient concentrations is tolerated as long as the minimum stoichiometric amount necessary for biomass production is provided by the culture medium. Nevertheless, a favourable C/N/P-ratio must be determined experimentally for each strain. Furthermore, magnesium concentrations tend to be critical if not maintained between 100% and 200% of the minimum demand in the culture medium. As for osmotic pressure and ionic strength, most microorganism cultures tolerate values between 0.1 and 10% solutes by weight[31].

1.2.3 Methanol

Methanol concentrations as low as 5–10 ppm, if evenly distributed in the culture, are sufficient for reasonable productivity[32]. On the other hand, changes in strain metabolism have been observed at methanol concentrations above 200 ppm[2,4,21,19]. A strong influence of concentration was observed by Harrison for *Pseudomonas extorquens*[42]. A maximum yield of $Y = 0.4$ was observed only for very low methanol concentrations, dropping swiftly to $Y = 0.32$ at 0.3% methanol and $Y = 0.2$ at 1%

methanol in the culture broth. Provisions should be made to maintain an appropriate concentration range even in the most inaccessable region of a reactor.

1.2.4 Sterility

In general, laboratory-scale experiments are carried out under aseptic conditions. Sterile running was reported to be necessary for the production of *Pseudomonas methylotropha* by Byrom et al.[15] and for *Candida boidinii* by Cardini[12]. On the other hand, methanol metabolism allows parameters to be set as to prevent infection. This is achieved by using synthetic media, avoiding the production of extracellular organic substances, adjusting the appropriate pH and temperature, and applying methanol as a substrate. Production at high dilution rates and under substrate-limiting conditions is favourable. Harrison has reported a mixed culture having been maintained trouble free for extended periods with no sterility precautions[42].

1.2.5 Carbon Dioxide

Although it is believed that molecular carbon dioxide inevitably dissolved in the culture medium during the reaction is actively involved in cell metabolism, no limiting concentrations have been evaluated according to Babij.[7]

1.2.6 Cell Density

The cell density of a culture influences metabolic parameters: at infinitive dilution (i.e. zero-cell concentration) maximum growth rate with minimum production of secondary metabolites is commonly observed whereas in laboratory reactors, there is a drop in both the growth rate and yield coefficients at higher cell concentrations. This may be due to the production of extracellular substances or to the oxygen limitation at the cell membranes[76,89]. On the other hand, industrial-scale SCP-production with *Pseudomonas methylotropha* is carried out at cell concentrations of $30–50\ g\ l^{-1}$ [16] without adverse effects. A deviation from linearity in a biomass-yield diagram was observed for *Methylomonas clara*[10] above $10\ g\ l^{-1}$ in a laboratory-chemostat culture.

1.2.7 Temperature

Operating temperatures are 35–42 °C for bacteria and 30–32 °C for yeasts. Thermophilic strains are described by Harrison et al.[43], Cooney et al.[21] and Snedecor and Cooney[102]. These data were evaluated by wash-out experiments[66]. They show a reasonable tolerance range of ±0.5 °C.

2 Biological Requirements for the Reactor Design

When considering the lay-out of a suitable production-scale, a typical culture medium can be defined as follows[40]:

biomass concentration of culture medium (dry weight basis)	$10–50\ g\ l^{-1}$
water content of cells[31]	80–85%
volume concentration of biomass in culture	5–33%
mean distance between two neighbouring cells in the	

medium with 5 vol.% biomass concentration	1.18 × cell diameter
mean distance between two neighbouring cells in the medium with 33 vol.%	0.16 × cell diameter
maximum cell package density[46]	10–15% dry weight
cell diameter of methanol-utilizing bacteria	0.3–0.8 μm[86]
cell diameter of methanol-utilizing yeasts	3–10 μm[4].

One of the tasks of the reactor is to supply the microorganisms with substrates and to remove waste products from each individual cell. Methanol conversion into biomass requires oxygen as an additional substrate. The oxygen demand depends on the yield, the maintenance coefficient of the culture and the chemical composition of the biomass[49,52,1,54]. Provided that all methanol is transformed to either biomass or water and carbon dioxide, a given yield coefficient corresponds to both a defined oxygen consumption and heat generation, as can be deduced by stoichiometry and Hess' law. The specific minimum oxygen demand for a given bacterial species (*Pseudomonas methylotropha*)[71] was calculated by MacLennan to be 0.88 kg O_2 per kg methanol and 1.92 kg O_2 per kg SCP. The oxygen demand of yeast-based processes is higher, for instance 3.25 kg O_2 per kg SCP[74]. A general relationship between substrate yield and oxygen demand according to Hamer[46] and Reuss[90] is shown in Fig. 2. For the commercial SCP production, the highest possible yield determines the gross feasibility of a process. As Littlehailes[68] reports from pilot plant processes, biomass/methanol yields of between 0.45 and 0.5 can be achieved by bacteria. Their oxygen requirements are taken as a basis for the following considerations.

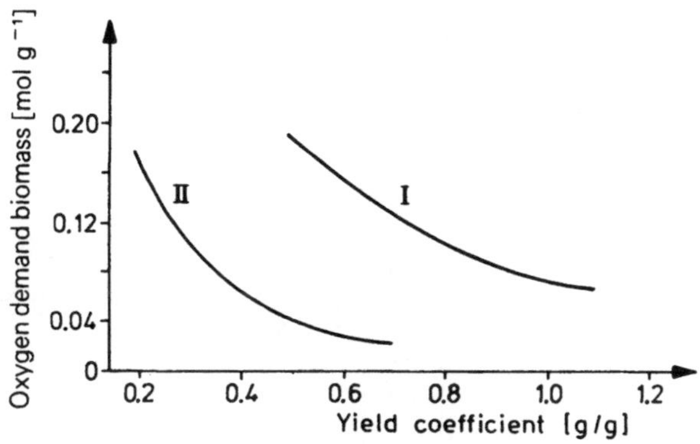

Fig. 2 Oxygen demand vs. yield coefficient (I: methane, II: methanol)

Combination of the oxygen requirements with the biomass productivities of 3–5 kg SCP per m^3 h[23] fixes the oxygen supply for any methanol utilizing strain at 6–10 kg O_2 per m^3 h, which generates a heat of reaction of 24.000 to 40.000 kcal m^{-3} h resp.[20]. Although it is desirable to have an economic oxygen uptake ratio in order to reduce the amount of air necessary for growth, 100% oxygen uptake from the air supplied cannot be achieved without simultaneous oxygen limitation of the culture which must be avoided. Data on pilot-scale reactors[38,74,40] indicate an oxygen yield of the reactor of 50% for an economic operation. This leads to an aeration rate of 0.8–1.3 vol^{-1} min^{-1}. Aeration rates of this magnitude give volumetric gas contents in the medium of up to 50%, depending on bubble size, viscosity, surface tension, and reactor design[87]. The mean Sauter bubble diameter can be calculated for $\varepsilon = 30\%$ and the gas-liquid interface $a = 1000\,m^2\,m^{-3}$ to be 1.8 mm. The mean distance between neighbouring bubbles is then $d = 0.15$–$0.2 \times$ bubble diameter. Based on the above mentioned data, a representative volume element of a cell suspension between two air bubbles is shown in Fig. 3. It follows that at any moment two neighbouring gas bubbles of a mean diameter of 1.8 mm are 0.4 mm apart. A static consideration would suggest that an oxygen molecule had to move a distance of at least 200 μm to the furthermost microbial cell, and even if it took the shortest possible route it had to pass 125 other cells each of which is acting as a mechanical obstacle and as an oxygen sink reducing its motility and average free diffusion path as compared to that valid in aqueous solutions. This abstract consideration of a static case has to be replaced by the consideration of the convective components of mass transfer created by the motion of bubbles relative to the liquid and by other mechanical mixing effects. By this motion, the static gas-liquid interfacial film is permanently renewed according to Danckwerts' liquid-film renewal theory[25]. Mass transfer calculations for the gas transport across a gas-liquid interphase usually take this interphase as being static and express both transfer velocities for the short time contact and the surface renewal effect by one transport coefficient k_L. This procedure incorporates, however, the danger of misinterpretation.

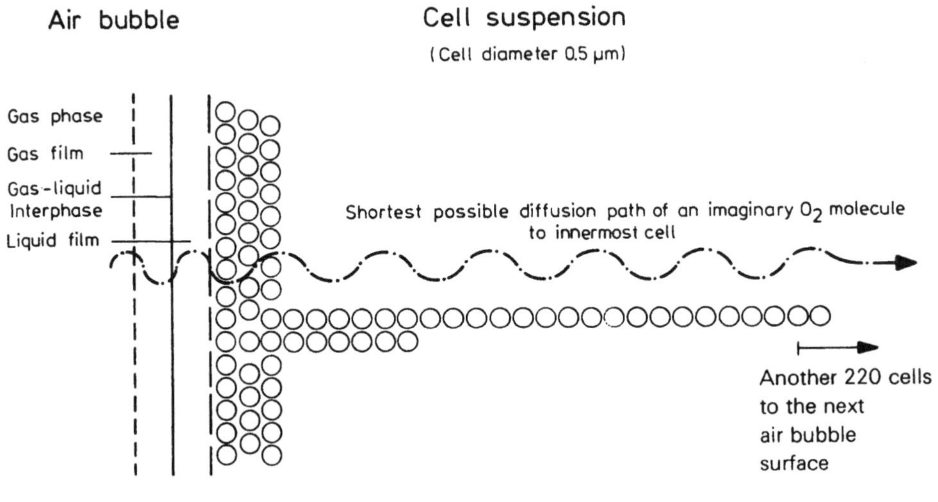

Fig. 3 Enlarged element of aerated culture medium

2.1 Specification of the SCP Reactor

Several authors have commented on the working volume of SCP-production. Lipinski comes to an annual economic production capacity of 100,000 tons SCP[67]. This implies a volume of 2500 m^3 at productivities of 5 g l^{-1} h for 8000 h annual running time. Even the production in parallel lines will require some hundred cubic meters of liquid volume. Blenke postulates a single large unit to be the most

economic one but, for safety reasons, suggests six 500 m^3 reactors for a 100,000-ton-per-year SCP plant[14]. Oldshue[80] reports 800 m^3 stirred tank reactors being built for the *n*-alkane SCP production in Southern Italy. Masuda et al.[72] quote one 750 m^3 reactor for a 50,000 ton per year plant based on ethanol.
Generally, three mixing principles are used in submerged process[85]:
Hydrostatic mixing (air lift);
mechanical mixing (stirrer, propeller; circulating pump);
hydrodynamic mixing (injector, ejector)
As mentioned earlier, biological requirements connected with the geometry and construction of a production are:
Easy and even distribution of all components throughout the cell suspension (gaseous and liquid substrates, microorganisms);
good temperature constancy by means of appropriate mass and heat transfer;
high biomass yield.
In addition, there are technical prerogatives to be fulfilled by a production:
Energy saving performance;
feasibility to continuous operation;
prevention of foaming;
production-scale realization;
aseptic performance.

2.2 Periphery of a Bioreactor

The periphery includes all parts of the plant which are not genuine parts of the reactor such as the piping for gas and substrates, analytical and control equipment, storage vessels containing the reactants, pumps and external heat exchangers. Although their connection with the reaction vessel proper depends on its construction and operation, these elements are employed as is usual in chemical engineering. Special care is taken to guarantee aseptic running and to avoid dead zones. Special lay-out investigations concerned with such details should be dealt with elsenwhere.

Often, a modern reactor is connected with an electronic data compilation unit in addition to its control equipment. An efficient operation control of equipment and calculation of process data (productivity, yield, mass balances) by electronic means has been established in fermentation plants. Full automatic control of a continuous reaction process seems to be inappropriate at present, because the tolerances and realiability of analytical probes as well as their response times are not computable in most cases[48].

2.3 Types Proposed

So far, the stirred tank reactor has been the conventional answer to aerobic requirements[59]. Katinger[56] points out that in order to increase the production scale, one should investigate small-scale equipment with deliberate heterogeneity to simulate conditions present in large reactors. Einsele[31] describes the poorer homogeneity as an increase of mixing times with the scale-up of stirred tank reactors. A comparison of different reactor types[101] is depicted in Figs. 4 and 5. A review of reactor principles available is given by Prokop and Votruba[85].

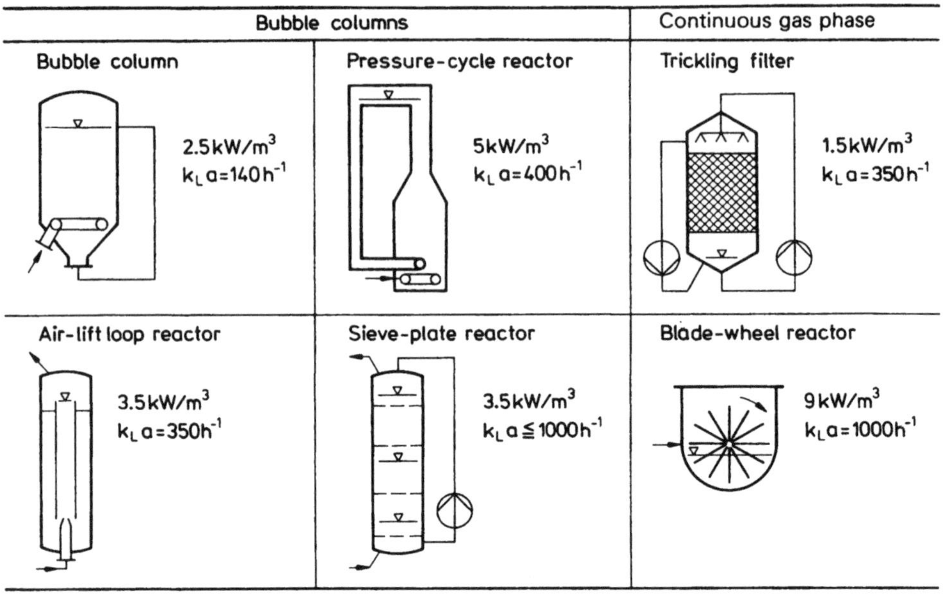

Fig. 4 SCP Reactors

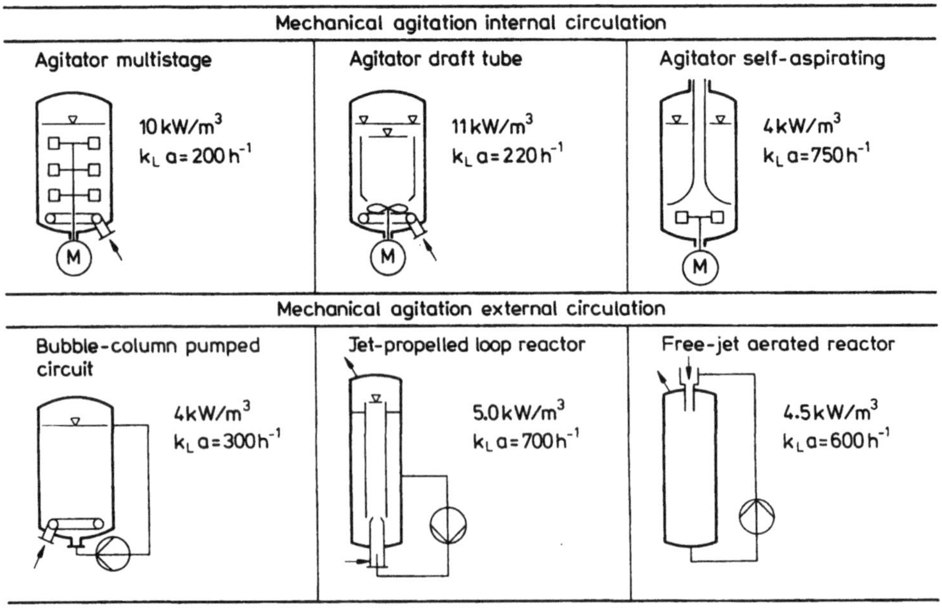

Fig. 5 Reactors (continued)

Sukatsch and Fritsch[104] have reported data on 4000 l stirred tank reactors for the protein synthesis based on *n*-paraffins:

Power input P/V [kW m^{-3}]	Biomass productivity Pr [kg m^{-3} h^{-1}]	Mass transfer coefficient (*ka*) [h^{-1}]
8	0.8 kg	202
3.8	0.5 kg	104

Ebner[29] has reported the energy required for the self-priming aeration of an acetic acid process to be 1.2 kW m^{-3} which is necessary to establish a gas-liquid interface of 1000 m^2 m^{-3}. Steiner, Moser and Lafferty[103] measured a mixing time of 80 s for water in a 25 m^3 deep jet aeration system to achieve 10% of inhomogeneity whereas Schreier[113] gives a power oxygen input ratio of 0.45 kWh kg^{-1} O_2 for a similar 2000 m^3 reactor. Faust[34] has reported a productivity of 3 kg m^{-3} h^{-1} in *n*-paraffin reactions in a self-priming reactor of 4000 l volume. Knecht *et al.*[58] have found the loop reactor with a jet drive to be suitable for the reaction of immiscible substrates. Ziegler[111] et al. examined a plug-flow loop reactor consisting of a pipe of 2.5 cm of diameter and 8 m of length, working with a fluid circulation velocity of 2.5 m s^{-1} with pressure control. Sittig et al.[100] refer to *n*-paraffin reactions in a bubble column of 1 m diameter consuming only 30 to 50% of the power input of a stirred tank reactor of equal size. The addition of inert 3 mm-granules of higher density than the broth results in a further rise in efficiency. Proportionality of the gas transfer rate with the aeration rate was demonstrated in a trickling filter reactor by the same author. Seipenbusch et al.[97] optimized a jet loop reactor as a reaction vessel for the cultivation of yeast on paraffin varying the recirculation rates of the broth. Rates of 400 circulations per hour have been reached. Ovaskainen et al.[82] have pointed out the importance of the spectrum of bubble residence times; extended recirculation rates may lead to low oxygen transfer rates in spite of a high gas liquid interfacial area due to the recirculation of exhausted gas bubbles. It becomes apparent that, depending on the requirements of a particular microbiological system, each type of reactor principle does offer its own advantages.

2.4 Bubble-Column Reactors

For large-scale biomass production on methanol, all important criteria can be favourably met by a bubble-column reactor:

Due to stoichiometry, high aeration rates are necessary for unlimited oxygen transfer in methanol reactions. This suggests to make use of the energy of mixing and flow already inherent in the compressed air. Lücke, Oels and Schügerl[70] have shown that this can be performed with bubble columns in a more energy-efficient way compared to stirred tanks. A bubble column is the only type of reactor which may have no moving parts. This provides an advantage when aseptic running is needed. Furthermore, it can be scaled up to large units without severe construction- and lay-out limitations.

Any reactor for aerobic submerged reaction processes can in principle be regarded as a bubble column, upon which — varying from case to case — additional mixing and pumping devices are superimposed, because the aerating gas has to be brought into contact with the reaction liquid, i.e. "to be bubbled through" the column of liquid. Much work has been carried out on the use of bubble columns as 2-phase-containing apparatus, mostly with small-scale laboratory equipment. Concerning scale-up, many aspects have hitherto not been studied scientifically[109]. The lay-out of large-scale apparatus must be based on small-scale experiments. Many fundamental physical properties will change with increasing reactor size in a partly unpredictable manner, especially the dependence of the bubble size on the flow and the change of hydrostatic pressure concentration gradients within the medium[70,110]. and Freedman Davidson[36] found little agreement among published gas hold-up data. This is due to the fact that the apparently simple reactor design of bubble columns includes a rather complicated gas-liquid flow pattern, particularly in large-diameter columns. A randomly induced back-flow in these larger units can be reduced by installation of flow guides. This leads to the concept of a loop reactor.

2.5 Loop Reactors

The solid phase, i.e. the microorganism particles present in the reaction medium, cannot be dealt with by means of classical thermodynamics. It is necessary to study microscale-concentration ranges by a quantum statistical approach to the description of the actual state of a reaction system. In this context, the concept of the homogeneously mixed stirred tank does no longer hold. The actual reaction system consists of individual cells and their micro-environment. The properties of these systems in a reactor can be described if a clearly defined flow pattern exists so that most physical gradients in the direction of flow occur and act upon each component of the system in the same predictable manner. These considerations lead to a concept which combines the advantages of bubble columns (i.e. simple design, lack of moving parts, economic energy consumption) with those of a loop reactor (i.e. defined mixing and circulation time distribution).

Several authors have reported that the loop reactor is superior to the classical bubble column:

Kanazawa[55] gives supporting evidence from an *n*-paraffin — based yeast process, Kuraishi et al.[60] observed a significant increase in $k_L a$-values by the use of draught tubes measured by sulfite oxidations. Huang et al.[53] introduced a kinetic model for the loop reactor which assumes plug flow in the annulus with a cascade of completely mixed and stirred tank reactors representing the draught tube section. His experimental data on 5 l and 32 l reactors are not, however, in all cases true for large units. Towell and Ackermann[105] examined the axial mixing of liquids and gases in bubble columns with a diameter of 40 cm and 106 cm, respectively. They found that liquid mixing was improved by a factor of 2.5 to 3 with the larger diameter and gas mixing was increased by the order of one magnitude when low gas velocities were used in the 106 cm column. The axial mixing coefficient of the liquid phase was doubled by installation of a draft tube.

Pollard and Shearer[84] compared an air lift reactor with a stirred tank. The air-lift reactor displayed a higher mass transfer with less energy consumption.

From these findings the following conclusions can be drawn:

A bubble-column type reactor is less energy wasting than a stirred tank reactor.

The flow pattern is markedly improved in bubble columns when flow guides or a draft tube are used. Bubble columns with flow guides meet all basic requirements for the SCP production in reactors summarized in Sect. 2.1. It is possible to aerate, degas or inject the various nutrients and substrates, remove heat and heat and harvest the product at any place of this loop system. Flow rates of the gas phase, the substrate and the cell suspension as well as the dilution rate under chemostat conditions may be adjusted both individually and independently (Fig. 6). In agreement with these considerations, most modern reactor systems display, to some extent, the loop principle.[24,28,85] Apart from the flat basin form which seems inappropriate for aseptic SCP reactions, two variations of vessel geometries are possible (Fig. 7) namely the compact, spherical vessel (H/D-ratio = 1) and the slim, cylindrical vessel (H/D ratio = 10).

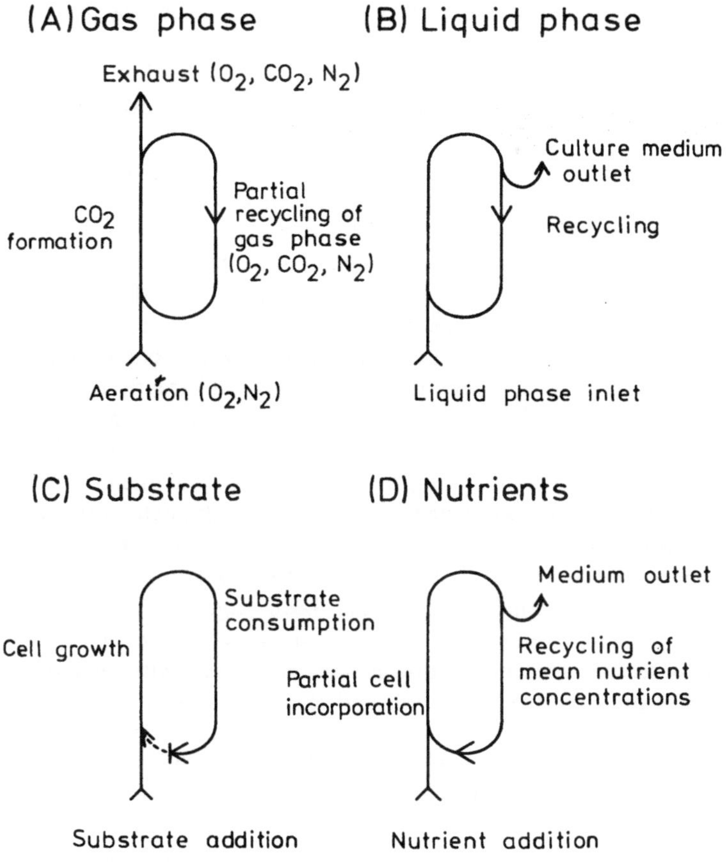

Fig. 6 The loop reactor principle

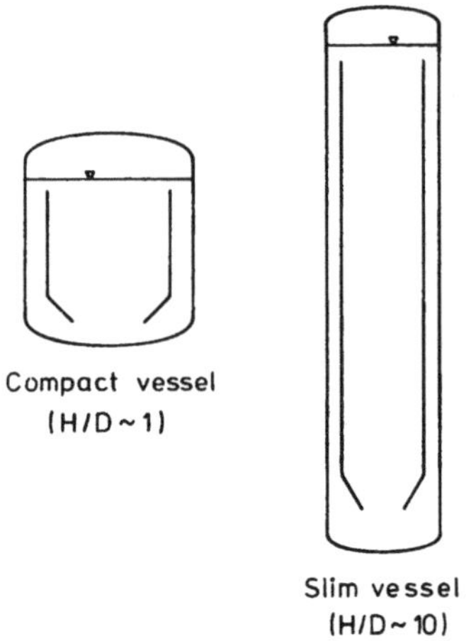

Fig. 7 Possible geometry of reactor vessels

In the case of the compact design, mechanical gas dissipation or hydrodynamic mixing and pumping become indispensable for high mass transfer. The air-lift drive of the aerating gas alone is too small because the reactor height necessary for a sufficient residence time is not provided even in large reactors. The scale-up is mostly empirical because of the increase of non-defined back- and radial-mixing with the enlargement of the reactor. The recirculation frequency cannot be predicted.

As for the slim reactor, an increasing reactor height increases the residence time of the gas phase and the mass transfer of oxygen and CO_2. In principle, additional stirrers, mixers or turbines can therefore be dispensed with. The loop system incorporates various additional advantages. There are no moving parts inside the reactor. Sterile running can easily be maintained for continuous operation. The circulation rate and mass transfer are sufficient at low specific energy demand. In their classical studies, Lefrançois and Revuz[63] compared different types of loop reactors all of which were run at high gas void fractions in the riser.

Blenke et al.[51,13] suggested cylindrical columns with an internal concentric draft tube which were originally applied to one-phase systems[11]. A minimum pressure drop was achieved in a one-phase system with a diameter ratio (D_t/D_A) of 0.59. With two-phase liquid-gas systems, the optimum diameter ratio can increase up to 0.9 according to Hatch[45]. The air-lift loop performance can be achieved as well by connecting two vertical tubes (riser and downcomer) at their tops and basis thus creating an external cycle, as has been suggested by Littlehailes[68], Pollard and Shearer[84] and Schügerl[95]. The reactors are characterized by a substantially smaller cross section area in the downcomer compared with that of the riser. This causes a pressure loss due to friction and less axial mixing. The advantage of an external loop is its capability to incorporate a simple external cooling device without the need of an additional mixing effect (Fig. 8): additional static mixing occurs at each circulation by division of the liquid stream when it leaves the tube and enters the annulus and its joining when it flows back into the draft tube. This principle already works for low circulation rates with pseudo-laminar flow. The aspects of the pressure drop are dealt with by Blenke et al.[11]. They show that for one-phase loop reactors, the

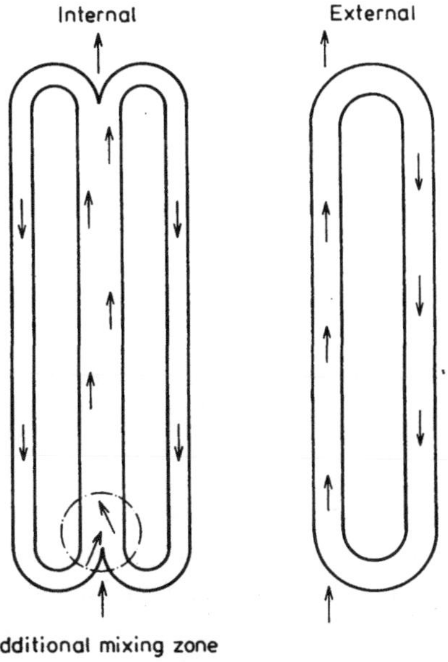

Fig. 8 Loop principle

ratio of the outer to the inner cross section areas can be adjusted to achieve a minimum pressure drop and a maximum flow. The shape of the reversion guides is important, streamlined deflectors preserving considerable amounts of kinetic energy. Hatch[45] has reported that for a given aeration rate, the liquid flow is increased with growing *H*/*D* ratio of the reactor vessel. In air-lift type reactors, the gas flow rate is the only source of energy input. An optimum cross section area ratio of riser and downcomer is given by Kosaric et al.[62] and Harrison et al.[43].

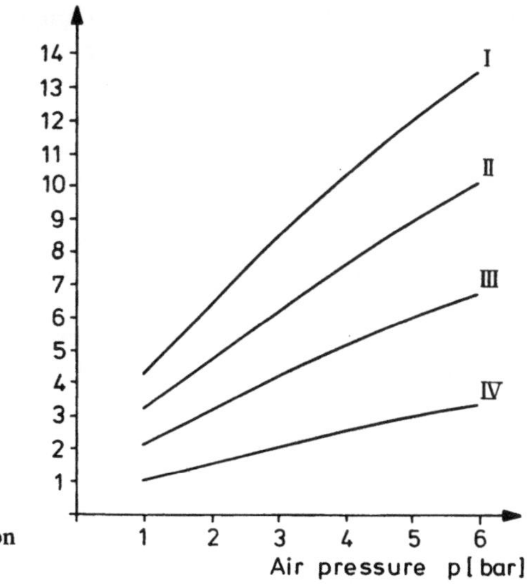

Fig. 9 Power demand vs. pressure head aeration rates [h^{-1}]: I 120; II 90; III 60; IV 30

Behringer[8] found that the air lift drive reaches a mechanical pumping efficiency of 50%. Fig. 9 shows the energy required for air compression in the range necessary for large-scale reactions[3]. According to the equation of polytrophic compression

$$P_{pol} = p_1 V_1 \frac{n}{n-1} \left[\frac{p_2}{p_1}\right]^{\frac{n-1}{n}}$$

reactors work more economically at higher pressures as long as there is a linearity between the oxygen transfer rate and gas pressure. Only reactors equipped with self-priming aerators can operate without precompression of gas in submerged cultures, as has been shown by Ebner[29] and Steiner et al.[103]. Therefore, the optimum working pressure for a given reactor heigh can be estimated. It is influenced by pressure- and oxygen tolerance of microorganisms;
economy of mass transfer;
optimum air lift drive;
stoichiometry and oxygen demand of the process.
It is understood that in large reactors with a large H/D ratio and with a high superficial air velocity, the predispersion of air by aeration devices has no effect on mass transfer. After an equilibrium has been attained in the reactor, the bubble size and surface area are determined only by liquid properties and by the turbulence reached according to Oels et al.[77]. The value of mass transfer in big stirred tank reactors largely depends on the kind of stirrer applied. This gives a power function as is shown by Pollard and Shearer[84] and by Hassan and Robinson[44].

The relationship between $k_L a$ and the gas power has been evaluated for bubble columns. It is clear that the amount of energy transferred to a bubble column is determined by the gas flow and its compression ratio. Oels et al.[77] as well as other authors have suggested the correlation

$$k_L a = b_1 (u_G/d_B)\, b_3$$

based on experimental work in small columns varying the superficial gas velocity. Its validity depends on the adaptation of the empirical constants b. Lücke et al.[70] observed an influence of the predispersion of the gas phase on the oxygen transfer coefficient of the reactor which was measured at low gas velocities, low aeration rates and in small columns only. Gasner[37] proposed an increase of $k_L a$ proportional to u_g, his work being based on a 100 l thin channel air-lift reactor with rectangular cross section. Kuraishi et al.[60] examined bubble columns with draft tubes in series and sieve plates in between. They found a favourable effect of these installations on $k_L a$ due to the repeated break up of bubbles. The measurements were carried out with a 450 mm column using the sulfite oxidation method i.e. for a non-coalescing medium. Towell and Ackermann[105], on the other hand, reported a decrease of axial dispersion by the installation of baffles.

2.6 Breaking of Foam

There is a special advantage of an internal loop for foaming reaction systems. As long as the tube in the centre of the column acts as a riser and, the annulus works as a downcomer; foam on top of the surface, which may attach to the wall, is washed down. As long as the liquid level above the tube is kept low enough to prevent the build-up of a liquid bubble column (Fig. 10) which does not take part in the general circulation, all foam is washed off the wall and redispersed into the broth.

2.7 Scale-Up of an Internal Loop Reactor

Most scale-up studies are obviously the concern of commercial companies and the findings of these studies are not available to the general audience. The scale-up procedure for a reactor to be applied to a particular process often involves preliminary calculations by an engineering team, test runs in a pilot plant, empirical optimization, and scale-up to the final production level. In the case of bubble loops, this procedure is generally based on the following assumptions (Fig. 10):

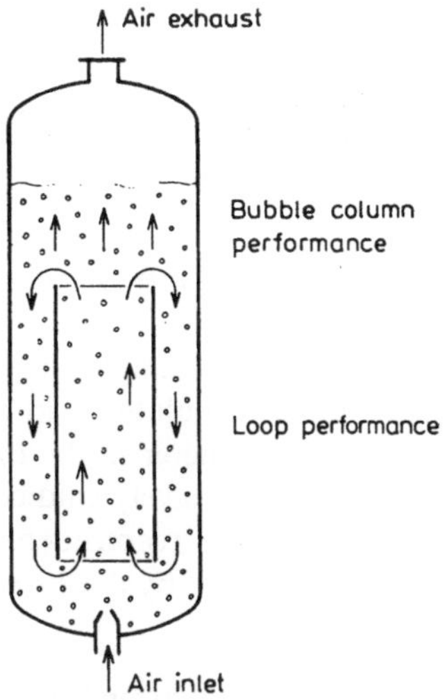

Fig. 10 Change of flow kinetics by high-liquid levels

The increase of the reactor volume with a similar reactor geometry results in higher $k_L a$-values and internal back-mixing[60, 105];
the increase of the superficial air velocity improves mass transfer rates and axial diffusivity (all authors);
the circulation can be altered over a wide range by choosing an appropriate riser-to-downcomer cross section ratio and installing baffles and screens;
a controlled back-mixing can be achieved by providing connections between the riser and downcomer (e.g. with perforations, slits or multiple tubes)[64].
According to Ovaskainen et al.[82] scale-up procedures concerning $k_L a$ are valid for one type of reactor only. Pollard and Shearer[84] propose a scale-up equation

for the diameter of a bubble column with almost stagnant liquid which is based on the critical gas content:

$$D = \left(\frac{4Q_G}{\pi W_s \varepsilon_{crit}}\right)^{1/2}.$$

$\varepsilon_{crit.}$ has to be estimated for each reactor to give an optimum mass transfer. It represents the gas void fraction when spouting performanc is reached.

Table 1. Scale-Up tendencies of Air-Loop Reactors

Physical feature	Tendency	
Gas turbulence	rising	$R_E \sim H/H_0$
Fluid velocity	rising	$\sim H/H_0$
Driving pressure	rising	$\Delta P \sim H/H_0 \times \varepsilon/\varepsilon_0$
Gas-hold-up	rising	$\varepsilon \sim (H/H_0)^N$
		$0 < N < 0.5$
Frictional pressure drop	declining	$\xi \sim \mathrm{Re}^{-0.237}$
Diffusive driving force	rising	$\Delta C \sim (H/H_0)$
Residence time of the gas phase	rising	$T \sim \varepsilon \sim (H/H_0)^N$
		$0 < N < 0.5$

3 Consideration of Mass Transfer

After the main dimensions of a reactor have been determined, lay-out parameters have to be calculated and adjusted to the particular process under consideration. The key parameter is the oxygen supply as stated by Reuss et al.[89)] and Miura[73)]. The latter extensively studied the theory of mass transfer in the reaction. Here, mass transfer will be reconsidered under the limitations set by a pneumatically driven loop-reactor and by the requirements of methanol processes as summarized above.

3.1 Oxygen-Transfer Resistance

According to Reuss et al.[89)], the oxygen molecules have to overcome the following resistances to reach the cells:
Diffusion from the centre of a gas bubble in the reaction medium into the inner gas boundary film;
diffusion through the gas film into the inner bubble surface;
transition through the gas-liquid interphase;
diffusion through the liquid boundary layer into the liquid;
transport through the liquid to the cell membrane;
transport through the cell membrane;
reaction resistances inside the cell.

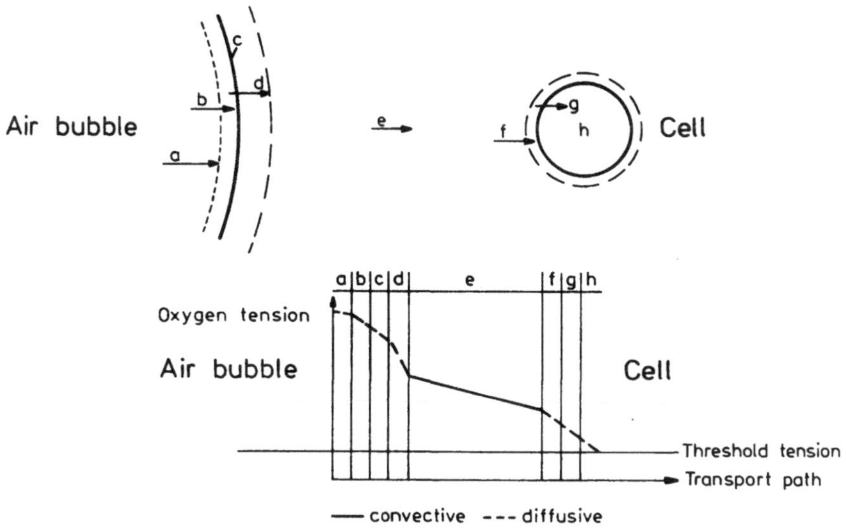

Fig. 11 Transfer resistance to oxygen from the bubble to the cell

The carbon dioxide produced by the metabolism has to leave the system along the reverse gradients. The corresponding transport equilibria and diffusion constants, however, differ from those of the oxygen transport, due to higher molecular solubility and different mobility. The other reactants to be brought to the cell surface from outside are water-soluble. The transport path of these reactants is shortened by the gas/liquid transition step. On the other hand, their even distribution in the cell suspension gains importance in production-scale reactors, since they will be introduced at one or a few places into the reactor only. From here, they must be mixed quickly with the culture medium in order to avoid local concentrations which might be toxic to the microorganisms. Primarily, however,the oxygen transport will be considered here.

Conventional hypotheses on oxygen transfer[88] postulate the liquid-film transition to be the rate-limiting step in the aerobic reaction. Taking into account the structure and physiology of the cell suspension it appears to be questionable whether measurements with the sodium sulfite system and with cell-free culture solutions can be used to describe the mass transfer in the multiphase biological system[106]. The sulfite/sulfate oxidation is modelled to be a first-order reaction depending on a given catalyst concentration only. It cannot be compared with the oxygen uptake by the metabolic paths of a living cell. Zlokarnik[112] proposed to define the quality of the medium as a coalescing or non-coalescent system. According to him and Schügerl[95], either the amount of the polar substance or the contents of ions present will determine this behaviour. Here, the oxygen is transported reversibly along numerous enzymatic reaction steps controlled by feed-back mechanisms.

The individual influences on the dynamics of oxygen transport, caused by convection, by reaction, and by pressure changes inside a reactor, are separately shown in Fig. 12:

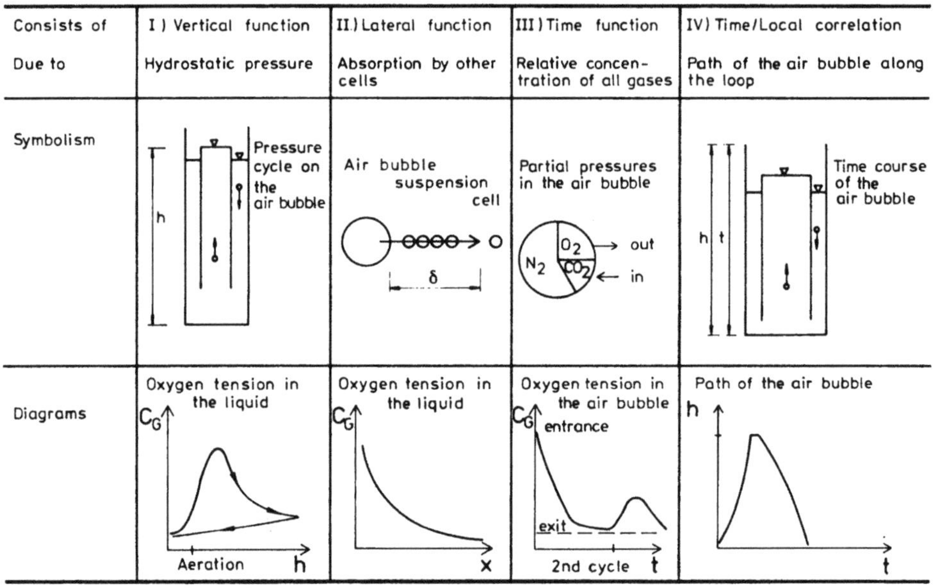

Fig. 12 Description of local oxygen tension in a loop reactor

3.2 Gas-Liquid Interface

In a tall loop reactor, an air bubble along its path from its formation at the bottom of the reactor to its disappearance at the top is subject to the following influences:
The hydrostatic pressure drop reduces the gas pressure inside the bubble;
the bubble expands and its surface area enlarges;
the bubble may coalesce with others reducing the specific surface area per gas volume and mixing the individual gas compositions;
the oxygen concentration decreases due to oxygen consumption by the culture;
the molar mass of the bubble decreases because of the 2:1 ratio μ of the oxygen absorbed to the CO_2 emitted (see Fig. 13); this, in turn, changing with the hydrostatic pressure;
the rising velocity of the air bubble is a function of the bubble diameter (see Fig. 13); this in turn, changing with the hydrostatic pressure.

In pure water, there is a maximum velocity at a definite bubble diameter, i.e. at the largest possible diameter of rigid spherical bubbles caused by the high surface tension of water according to Siemes[99)].

The bubble size of interest can be deduced from the phase surface area determined, for example, by the sulfite method[78)]. It follows that in methanol containing culture fluid, bubbles are formed which have a slip-rising velocity of 10–16 cm s^{-1} (Fig. 13). Because the measurements have to be carried out in stagnant colums at small gas flow rates to create single bubbles, the effects of turbulence and mutual drag are neglected. Therefore, these considerations are based on the data measured for pure water. In the case of high aeration rates, the bubbles do not rise as single entities but in swarms (Riquarts)[91)]. The relative rising velocity of the swarm is reduced

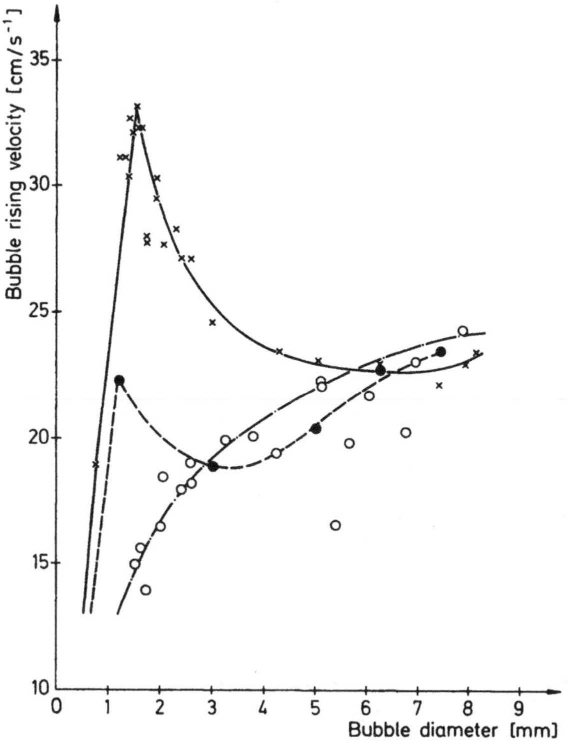

Fig. 13 Bubble rising velocity vs. bubble diameter

due to the fact that the liquid around each bubble has to flow back for continuity reasons thus creating an overall downstream of the liquid. Some authors (Lücke et al.[69], Deckwer et al.[110]) have studied the oxygen transfer in bubble columns and loop reactors. Their evidence does not cover large reactors. In particular, their gas velocities do not reach the range necessary for large reactors. Concentration measurements in the gas and liquid phases are often not reliable because the sampling of single bubbles cannot be carried out without disturbing the system. Schumm has developed an interesting sampling method of air bubbles in bubble-column reactors which may solve this problem[96].

Furthermore, oxygen electrodes allow only integral values to be determined and thus the actual concentrations around individual cells or air bubbles remain unknown. Microelectrodes developed by Tsao[106] have not yet been used in large vessels because of their mechanical fragility.

The most reliable experimental data are available for the gas balance. The following statements may be made:

According to Katinger[57] and Oels et al.[78], the region of the most intense oxygen transfer is the area of the bubble formation near the bottom of a reactor. The oxygen content of the air bubbles decreases as the latter rise, as does the oxygen partial pressure. Mass transfer in the upper region of a bubble-column loop reactor is diminished by low oxygen concentrations, low O_2 partial pressures and bubble coalescence. For small rigid bubbles, the oxygen concentration in the boundary region of a gas bubble compared with the oxygen concentration in the centre of a bubble is reduced due to diffusion. Apparently, along one radius the distribution of the oxygen tension in the gas phase has almost the same value because of the decreasing O_2 transfer velocity from a bubble along its rise to the surface. Direct mixing of gas through redispersion of bubbles as postulated by Seipenbusch et al.[97] is unlikely even in the case of additional mechanical acid. In general, therefore, the gas phase within a bubble reactor cannot be described to be completely mixed.

Particularly helpful is the volumetric factor B as defined by Blenke et al.[13]. It relates the film volume to the bubble volume. A small rising velocity of bubbles means a large factor B, i.e. the relative film volume to be penetrated is considerable. This film volume and hence B can be reduced by adjusting the hydrodynamics of the reactor, one of the tasks of reactor design.

3.3 Oxygen Balance in a Methanol Process

As has been shown for stirred tank reactors of production size by Oldshue et al.[81] and Blakebrough[9], considerable gradients of shear forces, gas contents and mixing intensity occur. A representative volume element of the culture medium is given in

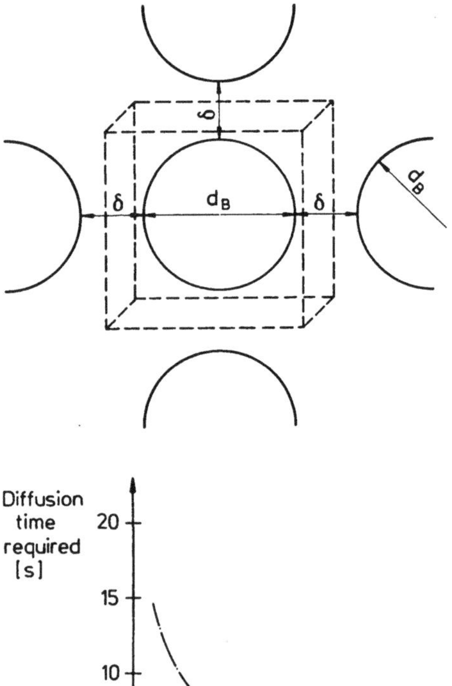

Fig. 14 Model of a representative volume element of a culture medium $d_B = 1.8$ mm, $\delta = 0.4$ mm, $a = 1000\ m^2\ m^{-3}$, $\varepsilon = 0.3$, $p = 1$ bar

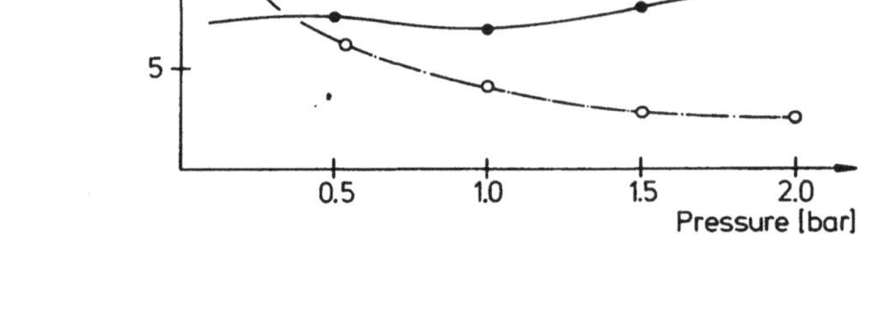

Fig. 15 Diffusion time vs. pressure
●—● 10% O_2 saturation $\delta = 200$ µm
○-·-○ 0.5 g/m³ O_2, $\delta = 200$ µm

Fig. 14. Fig. 15 is obtained after Danckwerts[25]. It demonstrates that the oxygen demand of the methanol system cannot be satisfied by diffusion alone. The plot shows the time required to reach a 10% saturation level in the 200 μm range around a stagnant liquid-bubble system. As soon as the bubble motion relative to the liquid is slowed down (e.g. in the case of a viscous medium and higher cell densities), the actual transfer comes close to this stagnant region if no other means of supporting microturbulence is employed. The maximum utilisable oxygen capacity of a liquid volume element is given by

$$m_{O_{2L}} = c_s - c_{min}\ (\mathrm{mg\ l^{-1}})$$

The following considerations are based on a loop reactor example with the lay-out data which are considered to be representative for a reactor of medium production size:

$H_L = 20$ m
$\varepsilon = 0.3$
$H_{total} = 29$ m
$D_A = 3.5$ m

$V_{total} = 300\ \mathrm{m^3}$
$V_L = 200\ \mathrm{m^3}$
$Q_{gas} = 200\ \mathrm{m^3\ min^{-1}}$.

According to Pirt[83] c_s amounts to 21 mg l^{-1} (35 °C) at the bottom of a reactor with 20 m of liquid height. At the head (moderate pressure) the saturation is 8 mg l^{-1} O_2 [75]. In most systems the average oxygen tension is aimed to be around 10% saturation as mentioned earlier. This results in a depletion interval of $m_{O_2}/Q_{O_2} = 0.5$ s. I.e. if full saturation could be achieved at the beginning of an experiment, the oxygen supply of the cells would cease after 5 s if not replenished.

The oxygen balance of the volume element of an aerated reactor can be expressed by the following equation.

$$V_B \frac{\partial c_B}{\partial t} = V_L k_L a (c_B - c_K)\,.$$

In case of oxygen limitation $c_K \ll c_B$. $K_L \cdot a$ is taken to be constant. Integration leads to

$$\ln (c_B - c_K) = \frac{V_L}{V_B} k_L a t + K\,,$$

$$= \left(\frac{1-\varepsilon}{\varepsilon}\right) k_L a t + K\,,$$

for $t = 0$ is follows $c_B = c_\alpha$ at the gas inlet and

$K = \ln (c_\alpha - c_K)$

thus

$$\ln \left(\frac{c_B - c_K}{c_\alpha - c_K}\right) = \left(\frac{1-\varepsilon}{\varepsilon}\right) k_L a t\,.$$

The amount of oxygen in one single bubble is

$$m_{O_{2B}} = V_B c_B .$$

the total amount of oxygen present in all bubbles can be estimated as

$$m_{O_2} = \varepsilon m_{O_2 B} = V_G \bar{c}_B .$$

Related to one m³ of the liquid volume, this gives a total of oxygen present in the reactor of

$$\frac{m_{O_2}}{V_L} = \frac{V_G}{V_L}\bar{c}_B = \left(\frac{\varepsilon}{1-\varepsilon}\right)\bar{c}_B .$$

For 50% of oxygen utilization, a good estimate of the actual mean gas concentration should be somewhat below 75%, i.e. the calculation is performed with 60% of oxygen concentration.

Then for 20 m of liquid height and $\varepsilon = 0.3$, the oxygen content of all bubbles present in the culture medium is less than

$$m_{O_2}/V_L = 0.15 \text{ kg m}^{-3} .$$

3.4 Significance of Convection for Mass Transfer

Sufficient oxygen supply to the microbial cells can only be achieved through convective forces on the microscale as has been shown above. Macroscopic back-mixing alone which occurs in the classical stirred tank is not sufficient. This is supported by Pirt's[83)] observation that in most cultivations the growth rate diminished with increasing cell density. The liquid mixing quality of loop reactors lies in the range of that of stirred tanks[13)]. The lay-out of a loop reactor has to aim at a defined energy dissipation field which must be present in any place of the total reactor volume to enable ample mass transport within the cell suspension between neighbouring air bubbles. Experimental examination of this microconvective field, however, encounters great difficulties. On the one hand, convection is achieved by the motion of each bubble relative to the liquid-cell suspension. On the other hand, the guided flow of the culture broth as it occurs in a loop reactor causes bulk turbulence which produces microeddies[61)]. Following Danckwerts[25)], the oxygen transfer due to convection, caused by the rising of gas bubbles, may be calculated as follows:

$$n_K = \frac{w_B d_B \pi}{A_B} = \frac{w_B}{d_B} (\text{s}^{-1}) .$$

For $d_B = 1.8$ mm and $w_B = 18$ cm s^{-1} n_k amounts to 100 s^{-1}. Each short-time contact causes an oxygen concentration drop of

$$\frac{\Delta c}{c_B} = \frac{4}{d_B}\sqrt{\frac{Dt_K}{2\pi}} .$$

(This assumption implies that the bubble can be regarded as a liquid sphere of constant diameter the oxygen concentration of which is equal to the equilibrium concentration of the interphase).

The relative rising velocity of bubble clusters lies in the range of 5–8 cm s^{-1} [77] whereas each individual bubble would rise faster, namely at 16 cm s^{-1} with an assumed bubble diameter of $d_B = 1.8$ mm in the low viscosity medium. The following calculation is based on this velocity which defines the frequency of the interphase renewal. The reaction rate of oxygen, as fixed by the demand of the microbial culture, must be met by the transfer rate (see Pirt[83]):

$$\mathrm{OTR} = Q_{O_2} \,.$$

All bubbles with equal diameters have a transfer capacity of

$$\mathrm{OTR} = \sum \frac{\Delta c}{c_B} c_B n_K \frac{1}{V_L} V_B = \frac{4}{\bar{d}_B} \sqrt{\frac{D t_K}{2\pi}}\, c_B \frac{1}{\bar{t}_K} \frac{\varepsilon}{1-\varepsilon} \,.$$

Insertion of the assumed values of a methanol cultivation into this equation leads to an oxygen transfer capacity, which is depicted in Fig. 16. The calculation is carried out in a stepwise manner.

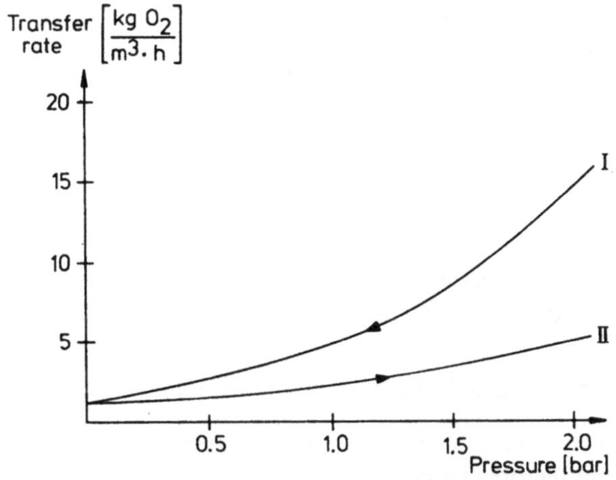

Fig. 16 Oxygen Transfer by short time contact I: riser; II: downcomer

The resulting curve does not take into account that in the region of fresh aeration the differential velocities of the gas and liquid are much higher and that the oxygen content of the fresh bubbles corresponds to saturation (23 weight %).

Consequently, the conventional approach to mass transfer should be reconsidered. It is misleading to express the convective contribution of the bubble slip velocity and its surface renewal potential by one static term 'k_L'. This transfer coefficient should contain the above mentioned short-time contact value as well as a term for the frequency of the surface renewal based on the slip velocity or an equivalent term expressing the convective contribution to oxygen transfer. For the lay-out of a gas-liquid contactor, it is essential to know the bubble size distribution in the vessel, the residence time distribution related to the bubble size and the void fraction of the gas in the broth.

For a gas-liquid interface of $a = 1000\ m^2\ m^{-3}$ and a gas hold-up of 30%, the mean bubble content of the aerated suspension amounts to 10^8 bubbles per m^3. Per unit of time, the bubble interface contacts a liquid film area of $aw_B d_B^{-1}$. Inserting the representative values, this leads to a gas-liquid interface renewal rate of $aw_B d_B^{-1} = 100.000\ m^2\ m^{-3}\ s^{-1}$, which amounts to a liquid film area of $5 \times 10^6 \times m^2\ m^{-3}$ contacted during the mean residence time of 50 s of the gas phase. To give a rough idea of the effective reduction of the actual distance x_{eff} which has to be overcome by diffusion, that interface renewal rate must be considered which has occured until the dissolved oxygen content is used up, i.e. within 5 seconds. Since the diffusive transport in the wake of each bubble is a two-dimensional problem, for calculation of the effective reduction of the distance of transportation, proportionality to the halfth power of the renewal rate is proposed. It cuts down the effective diffusion distance to

$$x_{eff} = 200\ \mu m \times 500^{-0.5} = 8.5\ \mu m$$

for the farthest cell. Since cells are present all over that length, only part of the oxygen required must be transported along the full distance. For estimates, this reduced length will be considered by taking 50% of that value.

With

$$D_{eff} = 2.0 \times 10^{-9}\ m^2\ s^{-1}$$

the remaining diffusive exchange is achieved by the theoretical driving force of

$$\Delta c_{diff} = OTR \frac{0.5 \times X_{eff}}{D_{eff} \times a}$$

$$\Delta c_{diff} = 6\ mg\ O_2\ l^{-1}$$

along the normal axis of mass transfer. Although this calculation is based on rough estimates it demonstrates that the oxygen transport between the interphase and the cell membrane cannot be explained by molecular diffusion.

3.5 Oxygen Transfer in the Downcomer

The oxygen transfer in the riser and in the downcomer is influenced by the following factors (Table 2):

Relative gas content,
average bubble size,
average oxygen concentration,
gas residence-time distribution,
increasing hydrostatic pressure and decreasing oxygen concentration with downflow,
liquid velocity.

Compared with the conditions in the riser, there is no gas inlet zone with its intensive aeration and dispersion which enables an accelerated transfer rate. For the downcomer the relative gas content as calculated earlier is lower than for the riser. The average bubble size must be assumed to be smaller than in the riser, since in the head space of the reactor, mainly the larger bubbles leave the

Table 2. Comparison of factors influencing OTR inside and outside of the Draft Tube

	Inside	Outside
Relative gas content	34%	26%
Mean bubble diameter	larger not quantitatively determined (see Fig. 13)	smaller
Mean oxygen concentration	70% saturation	35% saturation
Gas velocity W_G	$W_L + W_B$	$W_L - W_B$
Convective mixing	high turbulence in aeration zone turbulence due to liquid velocity and turbulence due to differential velocity between gas and liquid	—

suspension whereas smaller bubbles are recirculated (Fig. 12). In addition, the pressure increase with growing depth will result in bubble compression. The average oxygen concentration is lower than the oxygen concentration in the exhaust air leaving the reactor, i.e. below 50% saturation. The increasing pressure causes a reduction of the specific interfacial area: $a/a_0 = (p_0/p)^{2/3}$ may serve as an estimate. The driving concentration difference increases linearly with hydrostatic pressure:

$$\mathrm{OTR} \sim (k_L a)_0 \left(\frac{p_0}{p}\right)^{2/3} (\bar{c}_B \bar{p}/p_0 - c_K) .$$

This relation contains some simplifications which render an integration inappropriate. Though, the concentration decrease is somewhat compensated by the pressure increase. The specific interface may be correlated to the mean pressure.
Then for $c_k \ll c_A$:

$$\mathrm{OTR}_A \sim (k_L a)_0 \left(\frac{\varepsilon_A}{\varepsilon_E}\right) \left(\frac{\bar{p}}{p_0}\right)^{1/3} \bar{c}_A - \mathrm{OTR}_F .$$

OTR_F stands for the additional oxygen transfer in the aeration zone.

For the given example, this leads to a reduction of the average OTR in the downcomer of about 70–80% (see Table 2).

3.6 Gas- and Liquid-Mixing

Compared to the reaction rate of the biological metabolism of the methanol oxidation, the gas phase in a loop reactor cannot be considered as being ideally mixed. Rather, there exists a definite gradient alongside the reactor axis which is influenced by the circulation rate of the cell suspension, by the pressure drop and by the superimposed oxygen consumption. The circulation frequency of the gas content — taken from the gas balance — is given by:

$$Z_u = \frac{W_{LA}}{2H} = \frac{4}{58\ \mathrm{s}} \approx 250\ (\mathrm{h}^{-1}).$$

The gas recirculation number n_{UG} thus amounts to:

$$n_{UG} = \frac{Q_{\mathrm{circulating}}}{Q_{\mathrm{throughput}}} \approx \frac{0.50\ \mathrm{m}^3\ \mathrm{s}}{0.166\ \mathrm{m}^3\ \mathrm{s}} = 3.0.$$

An ideally mixed gas phase, however, would demand a circulation number $n_{UG} \geqq 20$[64].

A consideration of mass transfer has been restricted so far to the oxygen transport. The understanding of the process and reactor design must take into account all reaction components. Assuming passive transport by diffusion, Cooney et al. have roughly studied the transport of ionic and molecular components across cell membranes[23] for selected examples. Contrary to a chemical reaction such as, the sulfite oxidation which is very often used as a model for the investigation of the oxygen transfer, the methanol respiration and C_1-fixation consist of a series of enzymatic reaction chains. For most reactants, there exist individual productivities and yield optima. Surpassing of these concentration ranges results in either limitation or inhibition of growth. This effect can be observed pronouncedly for the methanol concentration. According to Lehnert, the theory of the loop principle[65] allows a maximum variation of 0.2% after 1 min after spot injection of a tracer. The circulation rates required are 300 h^{-1}. This postulate is easily fulfilled in a large loop reactor.

All liquid and dissolved reactants are introduced at one or a few spots into the reactor. In order to avoid the build-up of local concentration peaks, the recycling liquid passing the place of injection must take up the concentrated substrate (methanol, ammonia, ionic nutrients) in at least the inverse concentration ratio: The desired free methanol concentration in the reactor sould be below 10 ppm. If pure methanol is introduced into the reactor, the flow ratio of the circulating liquid to the methanol input must exceed 10^5.

The methanol volume flow is 12.5 $\mathrm{l\ m^{-3}\ h^{-1}}$; therefore, the injected methanol must be absorbed by a total liquid volume passing at a rate of 1250 times per hour. This means that the methanol must be injected at 5 different places for the given circulation number of 300 h^{-1}.

4 Air-Lift Drive

The calculation of the gas and liquid flow in a loop reactor is based on geometry and continuity. The driving force of an air lift drive is represented by the density difference between the riser and downcomer section. It is in balance with the total pressure drop of the medium during one circulation. Pressure losses at the head and bottom are determined individually by the method described by Blenke et al.[11] for one-phase systems. Local conditions of a bubble-column loop reactor are depicted in

Fig. 17. The gas void fraction for the following calculation is taken to be $\varepsilon = 0.3$. The same applies to the following parameters:

aeration rate:	$Q_{G_0} = \frac{60\ m^3}{m^3\ h}$
mean pressure:	$\bar{p} = 1$ bar
mean liquid velocity in the downcomer:	$\bar{w}_A = 4\ m\ s^{-1}$
circulation rate:	$Z_u > 300\ h^{-1}$

Then the mean residence time is given by

$$t_{w_{mean}} = \left(\frac{\varepsilon}{1-\varepsilon}\right)\frac{\bar{p}}{p_0}\frac{1}{Q_{G0}} = \sim 48\ s$$

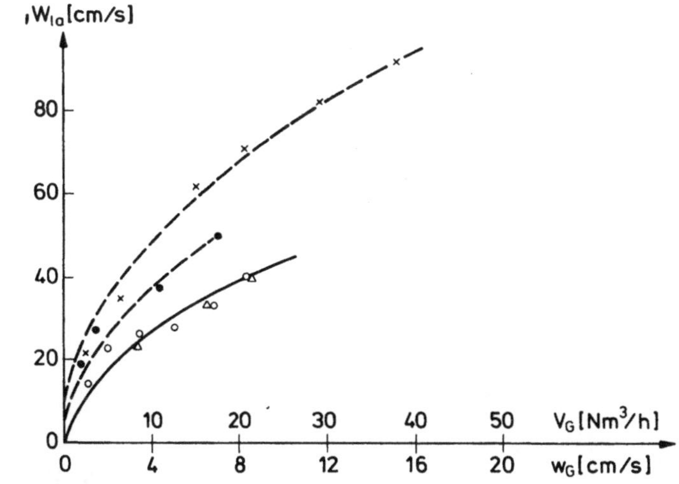

Fig. 17 Velocity of liquid in annulus, $H_L = 4.5$ m, $D_A = 0.3$ m

4.1 Gas Content in the Riser

The gas contents in the riser and downcomer can be determined by the following set of equations:

liquid balance:	$W_E F_E (1 - \varepsilon_E) = W_A F_A (1 - \varepsilon_A)$,
gas flow balance:	$(W_E + W_B) F_E \varepsilon_E = Q_\alpha + (W_A - W_B) \varepsilon_A F_A$,
volume constancy:	$(\varepsilon_E F_E + \varepsilon_A F_A) H = \varepsilon (F_E + F_A) H$.

This leads to

$$\varepsilon_E = \frac{Q + (W_A \varepsilon - W_B \varepsilon)(F_E + F_A)}{Q + (W_A - W_B \varepsilon)(F_E + F_A)}.$$

The driving density difference amounts to $\Delta\varrho = \Delta\varepsilon = \varepsilon_E - \varepsilon_A$. The values may be taken from Fig. 18.

$$\Delta p_{\text{drive}} = 2\,\frac{F_E H_E(1 - \varepsilon_E) + F_A H_A(1 - \varepsilon_A)}{F_E + F_A}\,\varrho_L\,.$$

This is the driving force of the air lift drive.

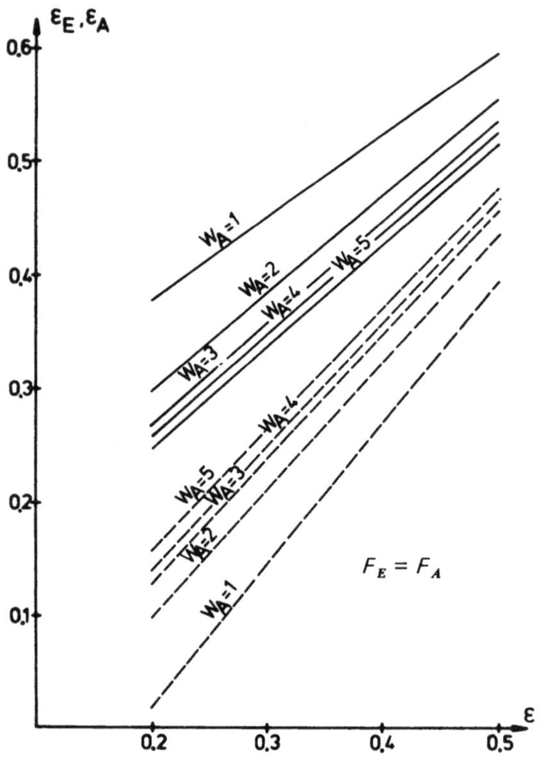

Fig. 18 Inner and outer gas content for a loop reactor

4.2 Pressure Losses

The coefficient of the decrease of pressure with viscosity can be taken as constant due to the high Reynolds number, as long as the viscosity is below 0.1 Pas. The reduction of the resistance due to the gas content of the culture broth is assumed to follow the factor $(1 - \varepsilon)$. For calculations, the head space is taken to be open. Which means that no external deflection of flow occurs, whereas the kinetic energy of the medium is destroyed through turbulent eddies. The pressure drop coefficients are

ξ_A = pressure drop of the downcomer
$\xi_{\cup}$ = pressure drop in the bottom deflection
ξ_E = pressure drop of the riser.

Hence, frictional losses are

$$\Delta p(\xi) = \left[\xi_U \frac{(W_E + W_A)^2}{2} + \xi_A W_A^2 + \xi_E W_E^2\right] \frac{\varrho_L}{2}(1 - \varepsilon),$$

$$\Delta p_{tot} = \Delta p(\xi) + \varrho_{L/2} W_E^2 (1 - \varepsilon).$$

The term $\varrho_{L/2} W_E^2$ stands for the pressure loss which is due to the kinetic energy of the liquid leaving the tube. Its energy is dissipated into small eddies.

For $F_E = F_A$, the equivalent hydraulic diameter of the annulus amounts to

$$D_h = D_A - (D_E + s) \approx 0.3 \times D_A .$$

This gives the following Re numbers:

$$\mathrm{Re}_A = \frac{W_A D_h}{\nu}$$

$$\mathrm{Re}_E = \frac{W D_E}{\nu} .$$

With $\nu = 2 \times 10^{-6}\ \mathrm{m^2\,s^{-1}}$ and $D_E = 0.7 D_A$ the pressure drop can be calculated according to[30]:

$$\xi = \frac{\lambda H_{tot}}{D} .$$

λ_E and λ_A may be taken from Ref.[30]. For the example of a loop reactor of 200 m^3 and 20 m of liquid height

$\xi_E = 0.107$ for the riser

$\xi_A = 0.29$ for the downcomer.

The pressure drop coefficient of the lower deflection can be taken from Blenke et al.[11]

$\xi_u = 3.0$.

The total pressure drop coefficient

$\xi_{tot} = \xi_U + \xi_E + \xi_A$

amounts to $\xi \approx 3.3$.

Hence, the toal pressure drop is

$$\Delta p(\xi) = 24{,}08\ \frac{\mathrm{kg}}{\mathrm{s^2\,m}} \approx 0.24\ \mathrm{bar},$$

$$\Delta p(\xi) = \varrho_W g[(1 - \varepsilon_A) - (1 - \varepsilon_E)]\, H_{tot},$$

$$\Delta p(\xi) = \varrho_W g H_L \frac{(\varepsilon_E - \varepsilon_A)}{1 - \varepsilon},$$

$\Delta p(\xi) = 0.224$ bar, as driving pressure ,

$\Delta p(\xi) = 0.24$ bar, as pressure drop .

The estimated liquid velocity of the example (a 20 m loop reactor of 200 m^3 liquid volume) must be corrected from 4 m s^{-1} to 3.86 m s^{-1}. The gas flow in the downcomer amounts to 0.5 m^3 m^{-2} s^{-1}.

5 Discussion

The scale-up of reactors from bench to production size significantly alters the physical parameters of the system. Characterization of conditions in the large scale by means of scale-up factors according to the similarity theory is not practicable, because a geometric similarity of the reactor vessel is not followed by a similarity of the physical properties in the reaction medium. Namely, the scale-up of tall reactors causes pronounced effects of pressure changes and accelerates the driving air-lift principle. Some conclusions for a further development of reactors may be drawn from these scale-up considerations and from the practical experience with processes described in literature.

It has been shown that the following parameters are increased with growing reactor height:

Superficial gas velocity,
air-lift drive,
liquid flow rate,
circulation rate,
concentration gradients of dissolved gases and spotwise introduced substrates,
gas void fraction

Consequently, the scale-up of bubble-column loop reactors leads to a better mass transfer. This performance is achieved with less specific energy input.

A direct equilibration of the gas composition between the individual gas bubbles is unlikely because the gas phase is not ideally mixed in a loop reactor. Therefore, it is necessary to consider a defined concentration pattern for the gas phase when scaling up a reactor. A critical point of investigation, though, should be the oxygen concentration in the downcomer. If necessary, an auxiliary aeration can be installed in this reactor.

The mixing intensity of the liquid phase in a large loop reactor as compared with that of the model is not expected to be inferior: the prolonged residence time of the gas phase causes an increase of the liquid flow rate with reactor height, resulting in a slightly accelerated recirculation rate. In one-phase loop reactors calculation methods for the quality of mixing and for recirculation rates have been published. For two-phase systems, the influence of the gas content on the overall mixing of the liquid has been determined experimentally for small-scale loops only.

It has been shown that for adequate substrate distribution, a recirculation number of 300 h^{-1} postulates the addition of methanol at more than four places. To secure degassing a higher circulation rate is not advisable. Thus, in large-production reactors, the level of the air inlet and hence the air compression ratio can be adjusted so as to save energy. For soluble substrates, the dissipation effect of an additional injector or ejector drive loses importance when the reactor is tall enough. Both air-lift and jet-driven systems are limited by the velocity of coalescence and degassing in the head space rather than by mass transfer. This velocity depends on the chemical composition of the medium and differs in the cases of pure water, water/methanol and water/paraffin systems. High aeration rates stoichiometrically necessary for the oxygen supply give sufficient gas velocities and interfacial areas without additional injector drive. A production-scale injector system creates a problem by the need for an external liquid cycle through a pump. For commercial pumps, there is a problem of building up the pressure necessary for the injector and working with a gas content of 30%. At the same time any preseparation of the cells to remove part of the air prolongs the residence time of the unaerated liquid and requires additional sterile equipment.

High aeration rates which are necessary for the SCP production render the large-scale application of stirred tank reactors uneconomic. The increase of the gas hold-up reduces the mixing efficiency of the stirrer. Compact stirred tanks of 8–15 m of height would have a gas velocity of 15–25 cm s^{-1}. The flooding point is surpassed at this velocity.

Furthermore, the loop reactor with its defined flow pattern and narrow local residence time distribution can enhance synchronous metabolism with its advantageous periodic pressure and concentration fluctuations[93, 107)].

The SCP production on methanol works at low viscosity in fully miscible nutrient media and high aeration rates. For this process, the loop reactor with a concentric inner riser is an efficient reactor design. Considering its properties, more questions about this reactor have been excavated than were answered.

7 Symbols

Symbol	Dimension	
A	m^2	surface
a	$m^2\,m^{-3}$	specific gas-liquid interface
a_0	$m^2\,m^{-3}$	specific gas-liquid interface at atmospheric pressure
B	$m^3\,m^{-3}$	volumetric relation of boundary film to bubble volume
BTM	kg	biomass on dry weight basis
c_B	kg m^{-3}	oxygen concentration in bubble
c_K	kg m^{-3}	oxygen concentration in liquid
c_α	kg m^{-3}	oxygen concentration at gas entry
D	h^{-1}	dilution rate
D	$m^2\,h^{-1}$	diffusion coefficient
D_{eff}	$m^2\,h^{-1}$	effective diffusion coefficient
D_h	m	hydraulic similarity diameter of annulus

Symbol	Dimension	
d_B	m	bubble diameter
D_A, D_E	m	reactor diameter
F	m^2	cross section of reactor
F_A	m^2	cross section of annulus
F_E	m^2	cross section of draft tube
H	m	height of reactor, or of liquid resp.
k_L	$m\ h^{-1}$	oxygen transfer coefficient
m_{O_2}	kg	mass of oxygen
n	—	exponent of polytrophic expansion
n_K	s^{-1}	frequency of contact
n_U	—	circulation number
P	kW	power
Pr	$kg\ m^{-3}\ h^{-1}$	productivity per volume
p_0	bar	atmospheric pressure
$\bar{p}$	bar	mean pressure in the reactor
p_{O_2}	bar	oxygen tension
$p(\xi)$	bar	pressure loss by friction
q	$m\ s^{-1}$	areal gas throughput
Q_G	$m^3\ h^{-1}$	gas throughput
Q_{O_2}	$kg\ h^{-1}$	oxygen uptake
OTR	$kg\ m^{-3}\ h^{-1}$	oxygen transfer rate
r	m	radius
Re	—	Re number
S	kg	substrate mass
s	m	wall-thickness of draft tube
t_D	h	doubling time
t	h	time
t_w	s	residence time
t_K	s	contact time
	—	oxygen turnover rate
U_G	$m s^{-1}$	superficial gas velocity after Oels
V_L	m^3	volume of liquid
V_B	m^3	bubble volume
V_R	m^3	volume of reactor
vvm	min^{-1}	volumetric aeration rate
w_B	$m\ s^{-1}$	slip velocity of a single bubble
w_L	$m\ s^{-1}$	overall gas velocity
w_E	$m\ s^{-1}$	liquid velocity in the draft tube
w_A	$m\ s^{-1}$	liquid velocity in the annulus
x	m	distance from the bubble surface
x	$kg\ m^{-3}$	cell density
Y	$kg\ kg^{-1}$	yield
Z_u	h^{-1}	circulation frequency
ε	—	gas content
ε_A	—	gas content of the annulus
ε_E	—	gas content of the draft tube
ν	$m^2\ s^{-1}$	kinematic viscosity
u	h^{-1}	specific growth rate
ξ_A	—	pressure drop coefficient in the annulus
ξ_U	—	pressure drop coefficient in deflection
ξ_E	—	pressure drop coefficient in the draft tube
λ	—	length-related pressure drop coefficient
ϱ	$kg\ m^{-3}$	density

8 References

1. Abbott, B. J., Clamen, A.: Biotechnol. Bioeng. *15*, 117 (1973)
2. Anthony, C.: J. Gen. Microbiol. *104*, 91 (1973)
3. Commercial prospectus: Aerzner Schraubenverdichter. Aerzner Maschinenfabrik, D 3251 Aerzen
4. Aiba, S., Humphrey, A. E., Millis, N. F.: Biochem. Eng., p. 22. New York: Academic Press 1973
5. *Anonymus*: Process Biochem. p. 30, Jan./Febr. 1977
6. Astuana, H. et al.: Biotechnol. Bioeng. *13*, 923 (1971)
7. Babij, T., Ralph, B. J., Rickard, P. A. D.: Proceedings of the Internat. Symp. Microbial Growth on C_1-Compounds, p. 213, Tokyo, Japan 1975
8. Behringer, H.: Die Flüssigkeitsförderung nach dem Prinzip der Mammutpumpe. Dissertation TH Karlsruhe 1930
9. Blakebrough, N.: Chem. Eng. *258*, 58 (1972)
10. Bender, R.: Hoechst AG, private communication 1976
11. Blenke, H., Bohner, K., Hirner, W.: Verfahrenstechnik *3*, 444 (1969)
12. Blenke, H., Prinzing, P.: Chem.-Ing.-Techn. *41*, 233 (1969)
13. Blenke, H., Hirner, W.: VDI-Berichte *218*, 549 (1974)
14. Blenke, H.: VDI-Berichte *277*, 127 (1977)
15. Brooks, J. D., Meers, J. L.: J. Gen. Mikrobiol. *77*, 513 (1973)
16. Byrom, D., Onsby, J. C.: I. Intersectional Congr. *77*, 513 (1974)
17. Cardini, F.: Yeast from Methanol. 5th Internat. Symp., Abstr., p. 202. Berlin: Verlag Versuchs- und Lehranstalt für Spiritusfabrikation 1976
18. Chalfan, Y., Mateles, R. I.: Appl. Microbiol. *135*, 23 (1972)
19. Colby, J., Zatman, L. J.: Biochem. J. *148*, 513 (1975)
20. Cooney, C. L.: Biotechnol. Bioeng. *11*, 269 (1969)
21. Cooney, C. L.: Ferm. Technol. Today *491*, 183 (1975)
22. Cooney, C. L., Levine, D. W.: SCP-production from Methanol by yeast, SCP II (S. R. Tannenbaum, D. I. C. Wang, eds.), MIT Press. London, 402 (1975)
23. Cooney, C. L., Levine, D. W.: Adv. Appl. Microbiol. *15*, 337 (1972)
24. Cooper, P. G., Silver, P. S.: Chem. Eng. Proc. *71*, 9 (1975)
25. Danckwerts, P. V.: Ind. Eng. Chem. *43*, 1460 (1951)
26. Van Dijken, J. P., Veenhuis, M., Harder, W.: Abstr. p. 387. V. Int. Ferm. Symp., Berlin 1976
27. Dostalek, M., Molin, N.: Studies of Biomass Production of Methanol Oxidizing Bacteria. SCP II: S. R. Tannenbaum, D. I. C. Wang (eds.), Chapter 19, p. 385. London: MIT Press 1975
28. Dunn, J., Blanch, H. W., Russell, T. W. F.: Chem. Rundschau *27*, 3.4, 17 (1974)
29. Ebner, H.: Abstr. p. 71. 3. Symp. Technol. Mikrobiol, Berlin 1973
30. Eck, B.: Techn. Strömungslehre. Berlin: Springer 1958
31. Einsele, A.: Abstr. p. 69. V. Internal. Ferm. Symp., Berlin 1976
32. Faust, U.: J. Ferm. Technol. *55*, 6 (1977)
33. Faust, U., Dorsemagen, B., Präve, P., Sukatsch, D. A., Zepf, Kh.: Abstr. p. 203. V. Internat. Ferm. Symp., Berlin 1976
34. Faust, U.: 1. Symp. Mikrobielle Proteingewinnung, p. 217. Braunschweig: Verlag Chemie 1975
35. Förstel, H., Schleser, G.: V. Internat. Ferm. Symp., p. 484, Berlin 1976
36. Freedman, W., Davidson, J. F.: Trans. Instn. Chem. Engrs. *47*, 251 (1969)
37. Gasner, L. L.: Biotechnol. Bioeng. *16*, 1179 (1974)
38. Giacobbe, F.: Economic Evaluation of New Trends in SCP Manufacture. ACS National Meeting, Philadelphia 1975
39. Goldberg, I.: Process Biochem. *11*, 12 (1977)
40. Gow, J. S. et al.: SCP II. In: S. R. Tannenbaum, D. I. C. Wang (eds.), Chapter 18, p. 370. London: MIT Press 1975
41. Häggström, L.: Appl. + Environm. Microbiol. *33*, 567 (1977)
42. Harrison, D. E. F., Wilkinson, T. G.: J. Appl. Batereol *36*, 309 (1973)

43. Harrison D. E. F., Topiwala, H. H., Hamer, G.: Proc. IV. IFS, Ferm. Technol. Today *491*, 000 (1972)
44. Hassan, I. T. M., Robinson, C. W.: AIChE J. *23*, 48 (1977)
45. Hatch, R. T.: In Single Cell Protein II, p. 46. In: S. R. Tannenbaum, D. I. C. Wang (eds.), p. 46. London: MIT Press 1975
46. Hamer, G., Topiwala, H. H., Harrison, D. E. F.: Erzeugung von Einzellerprotein aus Erdgas, gwf-gas/erdgas *114*, 531 (1973)
47. Harrison, D. E. F.: Making protein from methane. Chemtech. *9*, 572 (1976)
48. Heine, H.: Dechema Monogr. *71*, 53 (1954)
49. Herbert, D.: Stoichiometric Aspects of Microbial Growth. In: Cont. Culture, VI. Dean, Ellwood, Evans, Menning Ellis Harward Publish., Chichester 1976
50. Hirner, W.: Dissert. TU Stuttgart 1973
51. Hirner, W., Blenke, W.: Chem.-Ing.-Techn. *46*, 353 (1974)
52. Holve, W. A.: Process Biochem. 12 (1976)
53. Huang, S. Y., Yeh, M. C., Lion, K. T.: V. Internat. Ferm. Symp. Berlin 1976, Abstr. p. 68
54. ICI-Educational Publications, New Potein Appendix 1, The Kynoch Press, Birmingham
55. Kanazawa, M.: In: SCP II. S. R. Tannenbaum, D. I. C. Wang (eds.), p. 438. London: MIT Press 1975
56. Katinger, H. W. D.: Abstr., p. 72. V. Internat. Ferm. Symp. Berlin 1976
57. Katinger, H. W. D.: Berichte, Verlag Versuchs- und Lehranstalt für Gärungsgewerbe Berlin, p. 101, 3. Symp. Techn. Mikrobiologie, Berlin 1973
58. Knecht, R.: Process Biochem. *12*, 4 (1977)
59. Hubbart, D. W., Williams, C. N.: V. Internat. Ferm. Symp., Berlin 1976, Abstr. p. 71
60. Kuraishi, M., Matsuda, N.: Microbial Growth on C_1-Compounds, p. 231. Soc. Ferm. Techn., Japan 1975
61. Langmann, H., Taubert, C.: Verfahrenstechnik *2*, 10, 417 (1968)
62. Kosaric, N., Zajic, J. E.: Adv. Biochem. Eng. *3*, 89 (1974)
63. Lefrançois, L., Revuz, B.: Dechema-Monogr. *70*, 1327, 91 (1972)
64. Lehnert, J., Niewert, E.: Verfahrenstechnik *9*, 382 (1969)
65. Lehnert, J.: Verfahrenstechnik *2*, 58 (1972)
66. Levine, D. W., Cooney, C. L.: Appl. Microbiol. *26*, 6, 982 (1973)
67. Lipinski, E. S., Lichtfield, J. H.: Food Technol. 28 (5), 16 (1974)
68. Littlehailes, J. D.: 1. Symposium Mikrobiol. Proteingewinnung, p. 43. Weinheim: Verlag Chemie 1975
69. Lücke, J., Oels, K., Schügerl, K.: Dechema-Arbeitsausschuß, Technische Reaktion, Königstein, FRG 1976
70. Lücke, J., Oels, K., Schügerl, K.: Chem.-Ing.-Technik *49* (2), 161 (1977)
71. MacLennan, D. G.: Process Biochem. 6, *22* (1973)
72. Masuda, M., Nakauishi, Y., Sakakura, N.: Hydrocarbon Processing, p. 113, 1976
73. Miura, Y.: Adv. Biochem. Eng. *4*, 3 (1976)
74. Müller, F.: Abstr., p. 477. IV. Internat. Spec. Symp. on Yeasts, Berlin 1976
75. Näper, K. H.: Physikalische Chemie. Leipzig: VEB Deutscher Verlag für Grundstoffindustrie 1969
76. Naguib, M.: 1. Symp. Mikrobiol. Proteingew. p. 101, Verlag Chemie 1975
77. Oels, K., Schügerl, K.: Abstr. p. 64. V. Internat. Ferm. Symp. Berlin 1976
78. Oels, K., Lücke, J., Schügerl, K.: Chem.-Ing.-Techn. *49*, 1 (1977) MS 439/77
79. Ogata, K.: Agr. Biol. Chem. *33*, 1519 (1969)
80. Oldshue, J. Y.: Conf. Mass Transfer Scale- up Fermentations. New York: Henniker 1977
81. Oldshue, J. Y. et al.: Biotechnologie, Dechema, Tutzing Symp. 1977, *81*, Nr. 1678, Weinheim: Verlag Chemie 1977
82. Ovaskainen, T.: Process Biochem. *55*, 37 (1976)
83. Pirt, S. J.: Principles of Microbe and Cell Cultivation. Blackwell Scient. Publ., Oxford 1976
84. Pollard, R., Shearer, C. J.: Chem. Eng. *2*, 106 (1977)
85. Prokop, A., Votruba, J.: Folia Microbiol. *21*, 58 (1976)
86. Rehm, H. J.: Industrielle Mikrobiologie. p. 5. Berlin: Springer 1967
87. Reule, W., Schumm, W.: Inst. Chem. Verfahrenstechnik, Stuttgart, private communication (unpublished)

88. Reuss, M., Wagner, F.: 3. Symposium Techn. Mikrobiol., p. 89. Berlin: Verlag VLSF 1973
89. Reuss, M., Sahm, H., Wagner, F.: Chem.-Ing.-Techn. *16*, 669 (1974)
90. Reuss, M., Sahm, H., Wagner, F.: Chem.-Ing.-Techn. *46*, 16, 669 (1974)
91. Riquarts, R.: Chem.-Ing.-Techn. *49*, 1 (1977)
92. Rosenzweig, M.: Chem. Engng. *1*, 62 (1974)
93. Schleser, G., Förstel, H.: Abstr., p. 101. V. Internat. Ferm. Symp. Berlin 1976
94. Schlegel, H. G.: Allgemeine Mikrobiologie, p. 137. Stuttgart: Georg Thieme 1969
95. Schügerl, K.: Chem.-Ing.-Techn. *49* (8), 605 (1977)
96. Schumm, W.: Diplomarbeit Nr. 207e/th., Univers. Stuttgart 1976
97. Seipenbusch, R., Birckenstaedt, J. W., Schindler, F.: I. Symp. Mikrobiol. Proteingewinnung, p. 59, 1975
98. Seipenbusch, R.: Abstr., p. 56. V. Internat. Ferm. Symp. Berlin 1976
99. Siemes, W.: Chem.-Ing.-Techn. *26* (11), 614 (1954)
100. Sittig, W.: Abstr., p. 6. V. Internat. Ferm. Symp. Berlin 1976
101. Sittig, W., Heine, H.: Chem.-Ing.-Techn. *8*, 595 (1977)
102. Snedecor, B., Cooney, C. L.: Appl. Microbiol. *24*, 1112 (1974)
103. Steiner, W., Moser, A., Lafferty, R. M.: Abstr., p. 40. V. Internat. Ferm. Symp. Berlin 1976
104. Sukatsch, D. A., Fritsch, W.: Abstr., p. 53. 3. Symp. Technol. Microbiol. Berlin 1973
105. Towell, D. G., Ackermann, G. H.: V. Europ. Symp. Chem. React. Eng., Amsterdam 1972
106. Tsao, G. T., Lee, D. D.: AIChE J. *21*, 5, 979 (1975)
107. Tsao, G. T.: Abstr., p. 57. V. Internat. Ferm. Symp. Berlin 1976
108. Wagner, F.: Experientia *33*, 4, 110 (1977)
109. Wells, J., Giacobbe, F.: Commercial State of the Art of SCP, Internat. Board of R. & D. Conference, Washington, April 10, 1975
110. Zaidi, A., Deckwer, W. D., Adler, I.: Chem.-Ing.-Techn. *49* (6), 507 (1977)
111. Ziegler, H. et al.: Abstr., p. 66. V. Internat. Ferm. Symp. Berlin 1976
112. Zlokarnik, M.: Adv. Biochem. Eng. *8*, 133 (1978)
113. Schreier, K.: Chem.-Ztg. *99*, 7, 323 (1975)

Properties and mode of Action of Cellulase

Yong-Hyun Lee and L. T. Fan
Department of Chemical Engineering, Kansas State University,
Manhattan, Kansas 66506, U.S.A.

1 Introduction ... 102
2 Effect of Cellulase on some Properties of Cellulose Fibers ... 111
2.1 Fragmentation of Cellulose Fibers ... 112
2.2 Swelling of Cellulose Fibers and the Swelling Factor ... 113
3 Mode of Action of the Cellulase ... 114
3.1 ($C_1 - C_x$) Concept ... 114
3.2 β-1,4-Glucan Cellobiohydrolase ... 117
3.3 C_x Components ... 118
3.3.1 Endo-β-1,4-Glucanase ... 118
3.3.2 Carboxymethylcellulase (CMCase) ... 119
3.3.3 Exo-β-1,4-Glucan Glucohydrolase ... 120
3.4 β-Glucosidase ... 120
3.5 Synergism among Components ... 121
4 Physical Properties of Cellulase ... 124
4.1 Molecular Weights of Cellulase Components ... 124
4.2 Diffusion Coefficient and Sedimentation Constant of Cellulase ... 125
4.3 Size and Shape of Cellulase Molecules ... 125
5 Acknowledgement ... 126
6 References ... 127

Native cellulose undergoing an attack by cellulase exhibits extensive changes in physical properties prior to producing a measurable quantity of reducing sugar. These changes include fragmentation, swelling, considerable loss in tensile strength, transverse cracking, and lowering of the degree of polymerization. Fragmentation and swelling of cellulose fibers are characteristic of purified C_x enzyme action rather than C_1 action.

Reese and his co-workers (1950) first suggested a mechanism for the enzymatic breakdown of cellulose which involves a C_1 component. They postulated that the conversion of native cellulase was a two-step process: C_1 "activates" or desaggregates the cellulose chains, and the enzyme classified as C_x then carries out the hydrolysis reaction. Early research mostly related to studies of the C_x component. Since 1964, an extensive search for the C_1 component has been carried out, and substantial evidence was found to support the existence of a C_1-like components. Several investigators have supported the concept that the C_1 component has a nonhydrolytic function. However, based on the results of recent fractionation studies, it has been suggested that the ($C_1 - C_x$) concept be re-evaluated and that the mechanism of cellulase action be reformulated. The evidence cited by the various authors to support the claim that C_1 is a cellobiohydrolase appears to be convincing. Nevertheless, there remain striking differences in the extent to which highly ordered cellulose is hydrolyzed by the various C_1-type components.

The enzyme system of *Trichoderma sp.* has been examined more extensively than other enzyme systems. The mode of action of each component of this cellulase is shown graphically in Fig. 3 and can be summarized as follows:

Endo-β-1,4-glucanase. This contains several components with varying degrees of randomness. One of these may be the enzyme that acts on crystalline cellulose; however, it acts randomly, mainly on CMC, phosphoric acid-swollen cellulose, and cellodextrin. This component does not act on cellobiose. The main products are cellobiose and cellotriose.

Exo-β-1,4-glucanases. This is present in several forms. β-1,4-Glucan glucohydrolase removes a single glucose unit from the nonreducing end of the chain. This enzyme acts on Walseth, CMC, and cellodextrin chains of four to seven units produced by the action of the endo-glucanase but attacks insoluble cellulose with difficulty. This component has rarely been reported. β-1,4-Glucan cellobiohydrolase (CBH) removes a cellobiose unit from the nonreducing ends of the chain. The CBH is currently being equated with the classical C_1 enzyme by many investigators. This component has the greatest affinity for cellulose; it can not attack CMC and acts very slowly on H_3PO_4-swollen cellulose. Although it is unable to attack crystalline cellulose to any significant extent, it can degrade cellulose substrates by successively removing cellobiose residues from the chain ends. When CBH is recombined with C_x and β-glucosidase, it plays a major role in hydrolysis of cotton or Avicel-like crystalline cellulose.

β-Glucosidase. This hydrolyzes cellobiose and short chain cellooligosaccharides to glucose but has no effect on cellulose. While it rapidly hydrolyzes cellobiose and cellotriose, its rate of attack decreases markedly with an increasing degree of polymerization. This is in contrast to exo-β-1,4 glucanas, which acts preferentially on longer cellooligosaccharides.

The physical properties of each cellulase component, such as the molecular weight, the diffusion coefficient, the sedimentation constant, and the molecular size and shape, have been described herein. The molecular weight of the C_1 component from *Trichoderma sp.* lies in the range of 53,000 to 62,000. The endo-β-1,4-glucanases show considerable variation in molecular weights, ranging from 5,300 to 55,000. Exo-glucanase and β-glucosidase are reported to have molecular weights of 50,000 to 76,000. The molecular weight of β-1,4-glucan cellobiohydrolase was measured at about 42,000. If the cellulase molecules are spherical, their size would range from about 25 to 80 Å in diameter with an average of 60 Å. If the enzymes are ellipsoids with an axial ratio of about 6, their sizes would range from about 15 to 40 Å in width and from 80 to 250 Å in length, giving rise to an average size of 35 Å × 200 Å.

1 Introduction

Cellulase refers to a group of enzymes that contribute to the degradation of cellulose to glucose. The native crystalline cellulose is water insoluble, and its structure and complexity make it highly resistant to the hydrolytic action of enzymes. In most cellulolytic organisms, several cellulase components form a cellulase complex which synergistically hydrolyzes cellulosic substrates. Reese and his co-workers[84] have proposed the so-called $(C_1 - C_x)$ hypothesis to explain the mechanism of cellulase action. According to this hypothesis, the reaction involves at least two enzyme components. The first component, C_1, activates or deaggregates the cellulose chains in preparation for attack by the next hydrolytic component of the cellulase complex. The presence of component C_1 is essential if highly ordered substrates are to be attacked. In the next step, enzyme C_x hydrolyzes soluble derivatives of cellulose or swollen and partially degraded cellulose; however, highly ordered substrates are not attacked.

To understand the mode of action of a cellulase system, the substrate specificity of each of the cellulase components must be fully investigated after its purification. Prior to 1964, the microorganisms capable of extensive hydrolysis of the highly ordered forms of cellulose had not been screened. Also, techniques for fractionating cellulase components were not well-developed. As a result, most of the early investigations with cellulases were carried out using the C_x component. Several active C_x-type enzymes were found in culture filtrates of cellulolytic microorganisms. One component attacks cellulose endo-wise (randomly), and it exhibits different relative activities to different substrates. Other fractions act exo-wise (endwise) by removing successive units of glucose from the nonreducing end of a cellulose chain.

Recent studies of the mechanism of cellulase action have mainly been focused on the isolation, purification, and characterization of component C_1. Several investigators have supported the idea that component C_1 has a nonhydrolytic function, as originally proposed by Reese and his co-workers[84]. However, the results of recent fractionation studies suggest that component C_1 is not nonhydrolytic but a β-1,4-glucan cellobiohydrolase[5,10,24,118]. These investigators have proposed that the $(C_1 - C_x)$ hypothesis be abandoned and the mechanism of cellulase action be reformulated. According to their new concept, crystalline cellulose is effectively rendered soluble by the cooperative action of endo-glucanase and exo-glucanase enzymes; the latter removes cellobiose from the end of the cellulose chain. Many questions, however, still remain unanswered. Resolution of which depends on the development of purification techniques for cellulase fractions.

Comprehensive accounts of many aspects of cellulase and their applications are available in the literature[3,16,17,22,75,77,90,102,111]. The papers on cellulase presented at the 4th and 5th International Fermentation Symposia have been published collectively[9,96]. Articles on cellulase published before 1959 are well-reviewed in the book written by Gascoigne and Gascoigne[17]. A collective bibliography on cellulase and its application with 670 articles is available in the book edited by Reese[77]. Reese and Mandels[83] reviewed enzymatic degradation of cellulose. Nisizawa[46] also published a comprehensive review of the mode of action of cellulases. Summarized reviews of recent findings on enzyme mechanism associated with the C_1 component have been published[10,117].

In the previous work[15], the present authors have indicated that cellulose has a complex supermolecular structure and that the susceptibility of cellulose to such degradation is dependent on its structural features. This review will describe the enzymes that hydrolyze cellulose and related substances by focusing on their modes of action. Accumulated papers on the mode of action of cellulase, published since

1964, are listed in Table 1. These references are classified by research groups. Some of the references cited in Table 1 may appear to be obsolete or no longer relevant. However, all such references are included to illustrate, to the fullest extent possible, the pattern of historical development of various mechanisms to explain the mode of action of different kinds of cellulase and the relationships among the proposed mechanisms. Most cell free enzyme preparations from culture filtrates of cellulolytic microorganisms contain mainly the C_x component and β-glucosidase. Only a few cellulolytic microorganisms can produce a significant amount of the C_1 component.

Table 1. Mechanism and mode of action of cellulase

Authors	Microorganisms	Fractionation methods	Comments
Mandels and Reese (1964)	*T. viride*, *Myrothecium verrucaria*	Zone electrophoresis, DEAE-dextran chromatography	1) The hydrolytic (C_x) and solubilizing (C_1) factors of the cellulase complex were partially separated 2) Synergistic action of these two components on cotton digestion was observed
Selby and Maitland (1965)	*Myrothecium verrucaria*	Gel filtration on Sephadex G-75	1) Three major cellulolytic components were fractionated 2) One component was active on CMC degradation; the other two were mainly responsible for cotton degradation 3) There was no evidence of synergism between these components
Selby and Maitland (1967)	*T. viride*	Gel filtration on Sephadex G-75, Ion-exchange chromatography on DEAE and SE-Sephadex	1) Three components were fractionated 2) The components, which are essential for attack on cotton, were a CMCase, a cellobiase, and a third C_1 component which had no action on CMC, cellobiose, or cotton 3) Synergism was observed among these components
Selby (1969)	*T. viride*, *Penicillium funiculosum*	Gel filtration on Sephadex G-75, Ion-exchange chromatography on DEAE and SE-Sephadex	1) C_1 components, which have very similar properties, were isolated from *Trichoderma* and *Penicillium* 2) Synergism and cross-synergism between C_1 and C_x components were demonstrated for cotton degradation
Li, Flora, and King (1965)	*T. viride*	Chromatography on Avicel column	1) β-1,4-D-Glucan glucohydrolase (EC 3.2.1.21), β-1,4-D-glucan glucanohydrolase (3.2.1.4), and an enzyme component (C_1), capable of attacking crystalline cellulose, were purified 2) Synergism was observed among these components on cotton digestion 3) Distinct physical and enzymatic properties of each component were observed 4) C_1 component has a nonhydrolytic function

Authors	Microorganisms	Fractionation methods	Comments
Liu and King (1967)	*T. viride*	Not indicated	1) When Avicel was treated with partially purified C_1-cellulase, a marked increase in the total number of particles occurred during the initial stage of reaction 2) C_1 component may not be aw enzyme but rather a protein, which would cause the initial particles to gradually collapse and release the ultimate miscelles
Haliwell (1965)	*Trichoderma koningii*	Cell free filtrates	1) The early enzymic breakdown of cotton fibers is characterized by the formation of very short fibers
Halliwell and Riaz (1970)	*T. koningii*	Sephadex G-75, DEAE-Sephadex, CM-Sephadex, Cellulose powder column.	1) Two CMCase components and a C_1 component were separated 2) The C_1 component acted weakly and only on cotton, forming soluble products but not short fibers 3) The ability to form short fibers was confined to one of the CMCases, and its action was unaffected by the other components
Halliwell and Riaz (1971)	*T. koningii*	Similar to 1970 paper	1) Four apparently pure components, cellobiase, C_1-like component, CMCase, and that named component, C_2, were isolated 2) CMCase showed mainly short fiber-forming activity 3) Interaction of these four components in the degradation of native cellulose was discussed 4) Synergism was observed between C_2 and CMCase
Halliwell and Griffin (1973)	*T. koningii*	Recycling chromatography on DEAE-Sephadex and on Sephadex G-75	1) C_1 component was isolated, and it was shown to act as a β-1,4-glucan cellobiohydrolase 2) This component released terminal cellobiose units from cellulose. C_x component was not required for the action of C_1; however, the enzyme synergized extensively with cellobiase 3) The relationship between this C_1 component and the entire cellulase complex was discussed
Halliwell and Griffin (1974)	*T. koningii*	Not indicated	1) The activity of a highly purified C_1 component was shown to be that of an exocellulase, which liberates a cellobiose unit from •oth native and simple substrates. This reaction was promoted not by C_x but by cellobiase 2) C_x were composed of two components: CMCase and another component capable of hydrolyzing cellulose, but not CMC
Wood (1968)	*T. koningii*	Sephadex G-75, DEAE-Sephadex, SE-Sephadex	1) C_1 component was isolated 2) This component had little ability to produce soluble sugar from cotton; but when recombined with other components, this capacity was almost recovered 3) C_1 component had no swelling factor activity on its own, but it had a synergistic effect on the swelling factor activity with other components

Continuation of Table 1

Authors	Microorganisms	Fractionation methods	Comments
Wood (1969)	*Fusarium solani*, *T. koningii*	Ion exchange on DEAE-Sephadex, Gel chromatography on Sephadex G-100	1) C_1 components of *F. solani* and *T. koningii* were similar in physical properties 2) C_1 component of each strain synergized with the other strains C_x fraction (cross synergism)
Wood (1971)	*F. solani*	Heat treatment, Gel filteration on Sephadex G-100, Isoelectric focusing	1) The β-1,4-glucanase and the β-glucosidase components were purified 2) The specificity of these two components was studied
Wood (1972)	*T. koningii*, *F. solani*	Gel filtration on Sephadex G-75 or G-100, Ion-exchange chromatography on DEAE-Sephadex, Electrofocusing	1) C_1 component was separated. This C_1 component possessed a slight ability to produce reducing sugar from CMC 2) Strong synergism was observed between C_1 and C_x components 3) Cellobiose was the sole product of C_1 action on Walseth, dewaxed cotton, cellohexaose, and cellotetraose 4) C_1 is a cellobiosylhydrolase
Wood and McCrae (1972)	*T. koningii*	DEAE-Sephadex, with a salt gradient and pH gradient, Electrofocusing	1) C_1 component was isolated 2) This component had little ability to attack CMC and native cellulose but degraded Walseth cellulose readily and dewaxed cotton to the extent of 15% 3) C_1 component is a β-1,4-glucan cellobiosylhydrolase
Wood and McCrae (1975)	*T. koningii*	A series of fractionation and purification procedures	1) Eight highly purified components were purified, and all of them were found to be hydrolytic enzymes 2) C_1 acted synergistically to a great extent with more random acting C_x components
Wood and McCrae (1977)	*F. solani*	Ultrogel AcA-54, DEAE-Sephadex	1) The purified C_1 component showed little capacity for hydrolyzing highly ordered substrate but hydrolyzed readily Walseth soluble cellooligosaccharides 2) C_1 is cellobiohydrolase
Emert, Gum Jr., Lang, Liu, and Brown Jr. (1974)	*T. viride*	Avicel adsorption column, DEAE-Sephadex A-50, Preparative disc gel electrophoresis	1) Three forms of cellobiohydrolase and a cellobiase were purified 2) The function of the cellobiohydrolase is cleavage of cellooligosaccarides at the site of their production (i.e., the cellulose surface) 3) Inhibition of this component by the cellobiose was observed
Gum and Brown Jr. (1976)	*T. viride*	Ultrafiltration using Amicon PM 30, Avicel column, DEAE-Sephadex A-50	1) A glycoprotein enzyme β-1,4-D-glucan cellobiohydrolase (E.C. 3.2.1.91) was purified 2) Physiochemical properties were determined

Authors	Microorganisms	Fractionation methods	Comments
Pettersson, Axiö-Fredriksson, and Berghem (1972)	*T. viride*	Gel filtration on Bio-Gel P-10, Ion exchange on DEAE-Sephadex A-50, Preparative gel electrophoresis in polyacrylamide gel	1) C_1 component was isolated 2) C_1 component is a hydrolytic enzyme, it functions according to an endwise mechanism, possibly by removing cellobiose units 3) The synergistic effect with C_x enzymes was explained by supposing that the C_x enzymes randomly split β-1,4-glucosidic bonds and that the C_1 enzyme consecutively removes cellobiose units from the free end
Berghem and Pettersson (1973)	*T. viride*	Chromatography on Bio-Gel P-10, DEAE-Sephadex chromatography, Isoelectric focusing, chromatography on Bio-Gel P-60	1) Chemical and physicochemical properties of C_1 component were determined 2) Avicel, Walseth, and cellotetraose were degraded by the enzyme, and in each case the product was cellobiose 3) C_1 component is a β-1,4-glucan cellobiohydrolase
Pettersson (1975)	*T. viride*	Bio-Gel P-10, Ion-Exchange chromatography, Isoelectric focusing, Bio-Gel P-60	1) Four cellulolytic enzymes were purified 2) One of the enzymes was an exo-β-1,4-glucanase, which catalyzes the hydrolysis of microcrystalline cellulose up to 80% solubilization
Berghem, Pettersson, and Axiö-Fredriksson (1975)	*T. viride*	Same as 1972 paper	1) β-1,4-Glucan cellobiohydrolase was further characterized as regards chemical, physico-chemical, and enzyme properties 2) The enzyme is a β-1,4-glucan cellobiohydrolase 3) This enzyme can degrade microcrystalline cellulose up to 80% within 72 h if the cellobiose is continuously removed
Berghem, Pettersson and Axiö-Fredriksson (1976)	*T. viride*	Molecular sieve chromatography, Dipolar adsorbent chromatography, Isoelectric focusing, Affinity chromatography	1) A low molecular weight and a high molecular weight β-1,4-glucan glucanohydrolase (C_x) were isolated 2) Protein composition and molecular weight were determined 3) Both enzymes were active in releasing free fibers from filter paper, but low molecular weight enzyme was more effective
Eriksson and Rzedowski (1969)	*Chrysosporium lignorum*	DEAE-Sephadex A-50 column	1) Three cellulase peaks were separated 2) Strong indication for the existence of a C_1 enzyme was found
Eriksson and Pettersson (1975a)	*Sporotrichum pulverulentum*	Fractionation on DEAE-Sephadex and Gel filtration and SP-Sephadex, Concanavalin A-Sepharose, Polyacrylamide-	1) Five pure endo-β-1,4-glucanases were separated 2) One-exo-glucanase was identified 3) Endo-glucanases were purified and physicochemical characterizations were determined

Continuation of Table 1

Authors	Microorganisms	Fractionation methods	Comments
		gel electrophoresis, Isoelectric focusing on flatbed polyacrylamide gel	
Eriksson and Pettersson (1975b)	*Sporotrichum pulverulentum*	DEAE-Sephadex, Gel filtration, Activation on Dowex 2-X8 anion exchanger, Concanavalin-A-Sepharose, SP-Sephadex	1) An exo-β-1,4-glucanase was purified 2) Physicochemical properties were determined
Almin, Eriksson, and Pettersson (1975)	*Sporotrichum pulverulentum*	Same as Eriksson and Pettersson (1975)	1) Five types of purified endo-β-1,4-glucanase were characterized with regards to their molecular activities and Michaelis-Menten constants
Eriksson (1975)	*Sporotrichum pulverulentum*	Not indicated	1) The synergistic action among five types of endo-β-1,4-glucanases and an exo-β-1,4-glucanase was demonstrated 2) Quinone oxidoreductase was found
Iwasaki, Hayashi, and Funatsu (1964)	*T. koningii*	DEAE-Sephadex A-25, Amberlite IRC-50 (XE-64). Hydroxylapatite column	1) Two types of cellulase, cellulase I and II, were separated 2) Cellulase 1 had higher activity toward glycol cellulose and cellobiose but no measurable activity toward insoluble fibrous cellulosic material 3) Cellulase II could easily decompose fibrous cellulose, but did not act on glycol cellulose and cellobiose 4) Both enzymes were homogeneous and had different properties
Ogawa, and Toyama (1964)	*T. viride*, *Aspergillus niger*	Cellulose column using pulverized filter paper or disintegrated gauze or cellulose powder	1) Three cellulase components were proposed: C_1 (native cellulose hydrolyzing enzyme), C_2 (acting on filter paper or swollen cellulose) and C_3 (active on CMC) 2) Avicel was adopted as a new substrate for determination of activity
Ogawa and Toyama (1966)	*T. viride*, *A. niger*	Gauze column, DEAE-Sephadex A-50, Electrophorisis	1) Crude fractionation of cellulase from *T. viride* and *A. niger* was carried out 2) A cellulolytic component capable of degrading filter paper was separated 3) Cellulase components, which are active on Avicel, gauze, filter paper, and CMC, were studied
Ogawa and Toyama (1967)	*T. viride*	Gauze column, DEAE-Sephadex A-50	1) Four cellulase activities; the gauze degrading (C_1), Avicel decomposing (C_{1a}), filter paper degrading (C_2) and CMC decomposing (C_3) activities, were separated 2) Synergism among these components was observed

Authors	Microorganisms	Fractionation methods	Comments
Ogawa and Toyama (1968)	*T. viride*	Gauze column method	1) Cellulolytic enzymes were separated into non-adsorbed and adsorbed fractions by the gauze column method 2) The enzyme in the nonadsorbed fraction included cellobiase, CMCase, and Avicelase activity, but neither filter paper nor gauze degrading activities. The adsorbed fraction contained CMCase and Avicelase 3) Synergism was observed between these two fractions
Ogawa and Toyama (1972)	*T. viride*	Gauze column method, DEAE-Sephadex A-50, Amberlite CG-50, Sephadex G-100, Gel filtration	1) The nonadsorbed and adsorbed fractions were further purified 2) Nonadsorbed fraction was composed of two components and consisted of mainly CMCase activity. This fraction revealed no activities capable of degrading filter paper but showed trace activity against Avicel 3) The adsorbed fraction had three components which were active on filter paper, Avicel, and CMC. This fraction was considered to be an essential of natural cellulose decomposition
Watanabe (1968 a)	*Chaetomium globosum*	Cellulose powder column, DEAE-Sephadex A-25, Amberlite XE-64	1) Three cellulase fractions were separated 2) The enzyme system consisted of at least three fractions
Watanabe (1968 b)	*Chaetomium globosum*	Cellulose powder, DEAE-Sephadex A-25, Amberlite XE-64	1) One cellulase fraction, which acted on CMC, was separated 2) Properties of this fraction were determined
Niwa, Okada, Ishikawa, and Nisizawa (1964)	*T. viride*	Starch zone Electrophoresis, Hydroxylapatite, DEAE-Sephadex	1) Cellulase contained 3–4 isozymes; some exhibited strong CMCase activity, and others were active on filter paper 2) A synergism was observed among these components 3) Use of various substrates in measuring enzyme activity was suggested
Niwa, Kawamura, and Nisizawa (1965)	*T. viride*	Column chromatography using Amberlite, DEAE-Sephadex, Electrophoresis	1) Seven peaks of cellulase fractions were purified 2) One component was highly active on CMC, but much less active on insoluble substrates such as cotton and filter paper 3) One component, which showed homogeneous electrophoretic pattern, exhibited the C_1 and C_x activities 4) Further separation of the C_1 and C_x components was attempted but was not successful
Nisizawa, Suzuki, and Nisizawa (1966)	*T. viride*	Amberlite CG-50 column	1) Study on the nature and mechanism of swelling factor was performed 2) Swelling factor activity was probably involved in the cellulase activity itself 3) A semicrystalline region in absorbent cotton was suggested

Continuation of Table 1

Authors	Microorganisms	Fractionation methods	Comments
Okada, Niwa, Suzuki, and Nisizawa (1966)	*T. viride*	Amberlite CG-50, DEAE-Sephadex A-50	1) Five cellulase components were separated 2) One cellulase component hydrolyzed insoluble substrates, e.g., filter paper, more easily than CMC, cellodextrin and several cellooligosaccharides 3) It also showed a swelling factor activity toward absorbent cotton. Thus, this cellulase possesses C_1 as well as C_x activity with a random mechanism
Okada, Nisizawa, and Suzuki (1968)	*T. viride*	Amberlite CG-50 DEAE-Sephadex A-50	1) Three types of β-1,4-glucan glucanohydrolase were obtained 2) The properties of each component were examined 3) These types of cellulases hydrolyzed a series of substrates, including several cellooligosaccharides, CMC, and various insoluble celluloses such as Avicel and cotton, but they showed no activities toward cellobiose
Tota, Suzuki, and Nisizawa (1968)	*T. viride*	Amberlite CG-50, DEAE-Sephadex A-50, and Sephadex G-75 column chromatography	1) Kinetic constants of three purified cellulase components for various cellooligosaccharides were evaluated
Tomita, Suzuki, and Nisizawa (1968)	*T. viride*	Amberlite CG-50 Column chromatography, DEAE-Sephadex A-50, column chromatography	1) Chromatographic patterns of cellulase components of *T. viride* grown on the synthetic and natural media were compared with those of commerical product Meicelase and cellulase Onozuka
Nisizawa, Tomita, Kanda, Suzuki, and Wakabayashi (1972)	*T. viride*, *Irpex lacteus*	Amberlite CG-50 Column, DEAE-Sephadex A-50 column chromatography, Isoelectric focusing method	1) Two components, Avicelase and CMCase, were separated 2) Avicelase exhibited less random mechanism on attacking cotton fiber and CMC 3) CMCase reduced the degree of polymerization of cotton fiber more rapidly than Avicelase but produced a small amount of reducing sugar 4) Synergism was observed between two components of both strains 5) The authors supported the $C_1 - C_x$ concept but also proposed that their individual function should be modified
Tomita, Suzuki, and Nisizawa (1974)	*T. viride*	Amberlite CG-50 DEAE-Sephadex A-50 Biogel P-150	1) Further purification of Avicelase was attempted 2) Avicelase attacked Avicel very readily to produce a mixture of cellobiose and a few percent glucose, but attacked CMC less readily 3) Avicelase also attacked cotton and Walseth at a lower rate than Avicel 4) Avicelase was regarded as a cellulase component of the random type, but it was shown to be much less random

Authors	Microorganisms	Fractionation methods	Comments
Okada (1975)	*T. viride*	DEAE-Sephadex A-50, Sephadex G-100 Sephadex G-75 Starch column chromatography	1) Two cellulase components were separated 2) Some properties of these cellulases were investigated
Okada and Nisizawa (1975)	*T. viride*	Same as Okada (1975)	1) Two highly purified cellulases were obtained 2) Both cellulases produced predominantly cellobiose and glucose 3) Both cellulases were able to perform transglycosylations
Okada (1976)	*T. viride*	Similar to Okada (1975)	1) A cellulase component was purified 2) The enzyme was characterized as a less random type with regards to its action on CMC and produced predominantly cellobiose and glucose from various cellulosic substrates as well as from higher cellooligosaccharides
Kanda, Wakabayashi, and Nisizawa (1976a)	*Irpex lacteus*	Amberlite CG-50, Bio-gel P-100 gel, DEAE-Sephadex A-25, Sephadex G-100 gel, Starch column, Sephadex G-100 gel, rechromatography	1) An endo-cellulase was purified, and its physical properties were determined 2) This endo-cellulase attacked a series of cello-oligosaccharides, β-cellobioside, CMC, and insoluble cellulosic substrates 3) Cellobiose was the largest product from digestion of insoluble substrates
Kanda, Wakabayashi, and Nisizawa (1976b)	*I. lacteus*	Not indicated	1) Three endo-cellulase components were separated; one was CMCase, and the others were of the Avicelase group 2) A mixture of CMCase and Avicelase group gave a remarkable synergistic action in degradation of crystalline cellulose

They are mostly fungi such as *Trichoderma viride*, *Trichoderma koningii*, and *Fusarium solani*. Consequently, the mode of action of cellulase from these strains has been studied intensively. In addition to the mode of action of cellulase, physical properties of cellulase and the effect of cellulase on some properties of cellulose fibers are reviewed.

2 Effect of Cellulase on some Properties of Cellulose Fibers

Native cellulose undergoing attack by cellulase exhibits extensive changes in physical properties, such as transverse cracking, considerable loss in tensile strength, lowering of the degree of polymerization, increased capacity for moisture uptake and for alkali absorption, and fragmentation to short separable fibers, prior to producing a measurable quantity of reducing sugar[38,46,83,85]. These changes are associated with the insoluble nature of cellulose substrates and occur simultaneously or successively. The most significant phenomena are the fragmentation and swelling of native cellulose.

2.1 Fragmentation of Cellulose Fibers

In an early work, Marsh[44] reported that transverse cracks were observed to occur in cotton fibers upon a short exposure to cellulolytic enzymes, and he theorized that the enhanced fragility of cotton fibers may be due to these transverse cracks. Recently, notable investigations on the fragmentation of cellulose fibers have been carried out. Halliwell[23] reported that the early enzymic breakdown of cotton fibers is characterized by the formation of a large number of very short fibers that increases to a maximum and then decreases gradually by conversion to soluble sugars. The enzyme from *T. koningii* converted a minor fraction (up to 16%) of substrate into soluble sugar and a major portion (80%) into insoluble short fibers within 20 h. These fragments were converted further into the reducing sugar. Marsh[43] also reported that little production of soluble material occurred during the fragmentation of cotton fibers. The loss of fiber as nonsedimentable solid material did not exceed 6%, while 75% of the starting material was converted to shorter fibers during the treatment of dewaxed cotton fibers with a cellulase solution.

A similar observation on fragmentation was obtained by King[34,35], Rautela and King[74], and Liu and King[41] for various crystalline celluloses treated with a crude cellulase preparation from *T. viride*. Rautela and King[74] compared electronmicrographs of undegraded and 65% degraded crystalline cellulose I. The electron microscopic observation revealed that enzyme action resulted in fissures parallel to the long axes of the crystallite aggregates and production of many fine particles of similar shape and size.

Ogiwara and Arai[61] followed the attack of *T. viride* cellulase on bleached sulfite pulp and compared the action of the cellulase with the action of acid. Their work indicates that the cellulase preferentially attacks the amorphous regions of cellulose fibers as does the acid. However, they found the morphological character of cellulose fiber changes entirely upon treatment with acid. After a yield of residue of nearly 70%, both the length and width of fibers decrease. In contrast, very long fibers which have been only slightly reduced in width remain after treatment with cellulase, even at a residual yield of 40%.

Several researchers investigated which enzyme fraction is involved in this fragmentation. Based on the $(C_1 - C_x)$ hypothesis, the early stages of attack on cotton fibers can be tentatively identified with C_1 action[81]. The C_1 component of *T. viride* studied by Selby and Maitland[89] was also active in producing these short fibers. Another similar change has been studied by Liu and King[41], who found that the C_1 component would cause the initial particles to gradually collapse and release the ultimate micelles. Wood[112] also reported similar observations.

The fragmentation was studied in some depth by Halliwell and Riaz[26,27]. They separated two CMCase and two C_1 components. They found that the separated C_1

component acts weakly on cotton fibers, forming a soluble product, but does not contribute to short fiber formation. The short fiber-forming activity was attributed only to enzymes showing C_x activity, and their action was unaffected by the other components. In a more recent work, Wood[117)] further purified the enzyme fraction and found that fragmentation of Avicel is characteristic of purified C_x enzyme action rather than C_1 enzyme action. Pettersson[72)] also reported that the free fiber-forming activity is caused by endo-glucanase.

2.2 Swelling of Cellulose Fibers and the Swelling Factor

The most rapid and sensitive measurement of activity of cellulolytic enzymes is provided by the alkali swelling test. In an early work, Marsh et al.[45)] found that fibers invaded by some fungi or those treated with their culture filtrates enhance the absorbability of 18% NaOH. They assumed that this phenomenon is due to the action of an enzymelike entity produced by the fungi, and they named it the "swelling factor" (S-Factor). Later, Reese and Gilligan[81)] made a further study of this S-Factor in an attempt to fit it into the general action pattern of cellulase. They confirmed that its behaviour is that of an enzyme, concluded that the S-Factor could probably be included with the C_x enzymes, and suggested that swelling must occur as a result of the partial damage of the primary wall by the attack of the swelling factor secreted from the microorganism. This swelling occurs before release of reducing sugar, loss in weight, or decrease in the degree of polymerization.

On the other hand, Youatt and Jermyn[122)] and Youatt[121)] maintained that swelling is caused by the synergistic effect of several cellulases acting together to attack native cellulose. They also maintained that the swelling effect itself is in some way a consequence of the supermolecular architecture of a given fiber. Based on electron microscopic observation, Norkrans[55)] postulated that the extent of swelling after enzymatic attack may depend on the condition of fibrils, which constitute the network structure of the primary and the secondary layer in cotton fibers. A similar view was presented by Selby[85)], who concluded that the swelling of cotton fibers was enhanced by the loosening of the constricting effect of the winding layer during hydrolysis by cellulase.

Later Nisizawa et al.[51)] investigated the swelling factor of Meicelase, a commercial cellulase preparation from *T. viride*, using dewaxed absorbent cotton and mercerized cotton. They observed that these cellulose samples showed a similar response to the swelling factor, and the degree of swelling increased very rapidly especially at the early stage of incubation. The average degree of polymerization of cotton decreased in parallel with an increase in the extent of swelling during incubation with a cellulase preparation. In view of these results, Nisizawa et al.[51)] postulated that besides the crystalline and amorphous regions, there must exist another region in the microfibrils of cotton fibers, namely, a semicrystalline region with a moderate crystallinity. This region can easily absorb alkali even when only a few β-1,4-glucosidic bonds therein are cleaved by cellulase. However, they assumed that this region can not fully absorb alkali in the ordinary state but that following the attack by cellulase it might change temporarily to an amorphous state which can absorb more alkali.

3 Mode of Action of the Cellulase

3.1 ($C_1 - C_x$) Concept

Reese et al.[84)] first suggested a mechanism for the enzymatic breakdown of cellulose which involves the C_1 component. The conversion of native cellulose to soluble sugar was pictured to be a two-step process. They proposed that C_1 "activates" or desaggregates cellulose chains so that the enzymes classified as C_x can carry out the depolymerization as shown in Fig. 1. According to them, microorganisms that can grow only on soluble forms of cellulose, e.g., carboxymethylcellulose (CMC), synthesize only the C_x component, whereas microorganisms capable of growing on highly ordered forms of cellulose produce both C_1 and C_x. However this postulate of

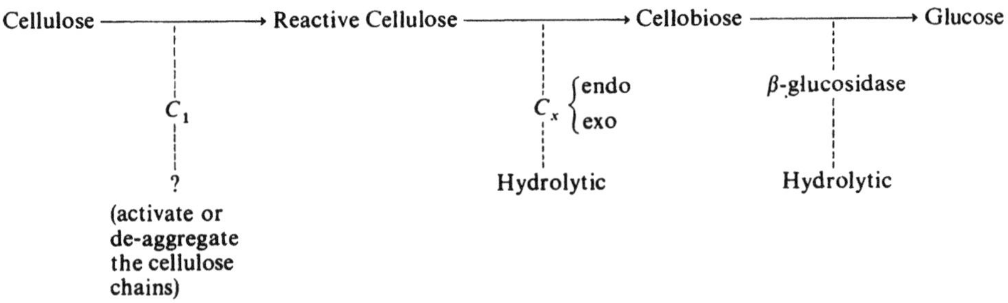

Fig. 1 Degradation of cellulose; the ($C_1 - C_x$) concept 42)

Reese *et al.* could not be confirmed. They, as well as other investigators, found that the cell-free culture filtrates show high activity only toward soluble, swollen, and partially degraded cellulose, which is hydrolyzed by the C_x enzymes. Because of the difficulty involved in preparing filtrates active against highly ordered forms of cellulose, e.g., native cotton, much of the work in the period, 1950–1964, was related to the studies of the C_x activity. However, the discovery in 1964–1965 that culture filtrates prepared from certain strains of *Trichoderma viride* and *Trichoderma koningii* are capable of extensive hydrolysis of native cellulose was a turning point in the study of cellulase. An extensive search for the C_1 component has followed. During this period, the separation of the C_1-like components from the culture filtrate of *T. viride* and *T. koningii* was reported by several groups of workers[23, 29, 40, 42, 53, 54, 56)].

Numerous reports dealing with the fractionation of the culture filtrates have appeared since 1964, and, in general, these have stated that cellulase, as secreted by *T. viride* and *T. koningii*, is a multicomponent enzyme system. Substantial evidence has been found to support the existence of the C_1 component; however, it appears that the majority of the investigators did not carry out the purification steps to completion, and thus, their views differ widely with regard to the specificities of the various C_1 enzymes that have been isolated. More recently, culture filtrates of *Fusarium solani*[113)], and *Penicillium funiculosum*[86)] have been shown to be highly active on cotton, and these have also been fractionated.

Several investigators have supported the idea that the C_1 component has a non-hydrolytic function. Mandels and Reese[42] have further developed the (C_1 — C_x) hypothesis on the basis of an observation that a culture filtrate of *T. viride* chromatographed on a DEAE-dextran column was separated into several components, including the C_1 and C_x components. Increases in the reducing sugar from cotton silver and CMC were used as measures for the activities of C_1 and C_x, respectively. It was suggested that cotton silver is first solubilized by C_1 and thereby changes into a form similar to that of swollen or generated cellulose, which, in turn, is hydrolyzed by C_x into soluble fragments. The C_x component shows no activity on cotton but does show activity on hydrolyzed amorphous cellulose as well as on CMC. Further purification of the C_1 and C_x components and the investigation of their properties have not been reported by these authors. The idea that the C_1 component has a non-hydrolytic function was supported by Li et al.[40].

Li et al.[40] purified three cellulase components: an enzyme removing successive glycosyl moieties from the nonreducing end, an enzyme attacking internal bonds, and a component capable of attacking crystalline cellulose. They postulated that the C_1 component has a nonhydrolytic function. Later Liu and King[41] observed that, when Avicel was treated with a partially purified C_1 component, a marked increase in the total number of particles occurred during the initial stage of reaction. They postulated that the C_1 component may no be an enzyme but rather a protein capable of causing the initial particles to collapse gradually and release the ultimate micelles.

Isolation of the C_1 component as an enzyme entity was reported by Selby and Maitland[88,89] and Selby[87]. This component was first found in the culture filtrates of *T. viride* and then in other sources such as *P. funiculosum*. The C_1 component becomes active only when the CMCase fraction from the same culture filtrate is added to it, but the C_1 component by itself shows no celluloytic capacity. Selby and Maitland[88,89] and Selby[87] concluded that this component represents a true C_1. Also, they postulated two possible explanations for the synergistic action of C_1 and C_x. The first points out that the C_1 component makes the cotton fiber more susceptible to the action of the C_x component. The second postulates that C_1 is inhibited by its own reaction products until they are removed by the action of C_x.

King and Vessal[36] have summarized the cellulase complex based on the (C_1–C_x) hypothesis as follows: C_1 is an enzyme whose action is unspecified, but it is required for the hydrolysis of highly oriented solid cellulose by β-1,4-glucanase (= C_x). C_x is a group of hydrolytic enzymes that exist as two types: (a) exo-β-1,4-glucanase, which successively removes individual glucose units from the nonreducing ends of cellulose chains, and (b) endo-β-1,4-glucanase, which randomly attacks internal linkages that are more susceptible to hydrolysis than terminal linkages. According to the (C_1–C_x) theory, cotton fibers are attacked successively by C_1, C_x, and β-glucosidase (cellobiase) and are finally converted into glucose. The β-1,4-glucanase and β-glucosidase will be discussed in more detail in the succeeding section.

On the other hand, some experimental results appear to contradict the (C_1–C_x) theory. The C_1 component, which is the essential component in hydrolyzing crystalline cellulose, has shown a hydrolytic rather than a nonhydrolytic function[14,

[26,47,58,59,60,100,112]. Ogawa and Toyama[56,59,60] reported on the separation of two cellulase components. One component consisted mainly of CMCase activity and revealed no activities capable of degrading filter paper but showed trace activity against Avicel. The other component was active on filter paper, Avicel, absorbent cotton, as well as CMC. Thus, these researchers considered the latter to be the essential component of the natural cellulose decomposing activity of *T. viride*. Wood[112] separated a C_1 component that had little ability to produce soluble sugar from cotton but when recombined with other components almost totally recovered this capacity. In addition, this C_1 component had no swelling factor activity on its own, but it had a synergistic effect on the swelling factor activity with other fractions. Strong indication for the existence of a C_1 enzyme from another strain, *Chrysosporium lignorum*, was also reported[14].

Experimental results by Okada et al.[65,66] also appear to contradict the $(C_1$–$C_x)$ theory. These investigators obtained three highly purified cellulase components from *T. viride*, each of them showing both C_1 and C_x activities; that is, they hydrolyzed randomly not only cotton fiber and Avicel but also CMC and various β-1,4-oligoglucosides excluding cellobiose, although the relative ratios of the activities of the three highly purified enzyme components for each of these substrates are considerably different. All purified cellulase components appear to be of the mixed $(C_1 - C_x)$ type, although the C_x activity in one component may be different from that in another, with each exhibiting a unique substrate specificity.

Recently Nisizawa et al.[52] further purified the cellulase complex and separated two different components, namely, Avicelase and CMCase. The purified CMCase easily and randomly saccharifies highly amorphous cellulose, CMC and cello-oligosaccharides. It produces glucose and cellobiose in almost equal amounts. However, it does not readily attack Avicel and cotton. In contrast, the purified Avicelase splits not only highly crystalline cellulose such as Avicel and cotton but also cellooligosaccharides to produce almost exclusively cellobiose from these substrates. The Avicelase also hydrolyzes CMC, but in a much less random fashion than CMCase. Therefore, the Avicelase seems to cut off one cellobiosyl residue at a time, mainly from the end of these substrates. Nisizawa et al.[52] also found that a mixture of CMCase and Avicelase shows a remarkable synergistic effect toward highly crystalline cellulose such as Avicel and cotton but not highly amorphous cellulose, such as phosphoric acid swollen cellulose[101] and CMC. A similar result was obtained from a culture filtrate of *Irpex lacteus*[52]. Thus, according to Nisizawa et al.[52], the CMCase and Avicelases from these fungi seemed to correspond to C_1 and C_x, but it must be CMCase that attacks cellulose initially to give cellulose fragments with low degrees of polymerization. These fragments, in turn, serve as substrates for Avicelase; this cellulase saccharifies the cellulose fragments from their ends. Thus, they suggested that the roles of C_1 and C_x are opposite to the roles originally proposed by Reese et al.[84]. They also reported that the separated C_1 component is a hydrolytic enzyme. This component functions according to an endo-wise mechanism, possibly by removing cellobiose units. The synergistic effect with the C_x enzyme was explained by supposing that the C_x enzymes randomly split β-1,4-glucosidic bonds and that the C_1 enzyme consecutively removes cellobiose units from the free end.

3.2 β-1,4-Glucan Cellobiohydrolase

Some investigators have supported the concept that the C_1 component has a non-hydrolytic function. However, based on the results of recent fractionation studies, several investigators have suggested that the concept be abandoned and that the mechanism of cellulase action be reformulated. The new concept postulates that crystalline cellulose is effectively rendered soluble by the cooperative action of an endo-glucanase enzyme and an exo-glucanase enzyme, which performs the special function of removing cellobiose from the end of the cellulose chain.

The properties and mode of action of an individual enzyme of the cellulase complex can be described with confidence only when its purification is complete. Chromatography on columns of Avicel, hydroxylapatite, Amberlite ion-exchange resins, alkali swollen cellulose, and degraded cotton gauze have been used to fractionate celluloytic enzymes (see Table 1). The most successful results have been obtained with fractionation using a combination of more than one packing material. Some researchers have effectively used a combined procedure of gel filtration and chromatography on various forms of ion-exchange Sephadex in the fractionation of the cellulase complex.

Wood[116)] and Wood and McCrae[118, 119)] reported on the isolation and purification of the C_1 component produced by *T. koningii* and *Fusarium solani*. The isolation and purification were carried out by gel filtration on Sephadex G-75 or G-100, ion exchange chromatography on DEAE-Sephadex, and electrofocusing in a stabilized pH gradient spanning over half a pH unit. The C_1 component obtained still possessed a limited ability to produce reducing sugar from a solution of CMC but showed no capacity for reducing the viscosity of the solution. It was concluded that C_1 activity and a trace of C_x (CMCase) activity originated in the same enzyme protein. Furthermore, the C_1 component, acting alone, was unable to attack highly ordered celluloses; acting synergistically with the separated C_x and β-glucosidase components, it was able to accomplish extensive breakdown. Cellobiose was virtually the sole product of the action of C_1 on (a) phosphoric acid-swollen cellulose, (b) dewaxed cotton fiber that had been 15% degraded by exposure to a highly concentrated enzyme for four weeks, and (c) cellohexaose and cellotetraose. Wood[116)] and Wood and McCrae[118, 119)] advocated that the C_1 component is a β-1,4-glucan cellobiosylhydrolase. Wood[117)] has formulated that the C_1 is a special type of hydrolytic enzyme which is unable to attack CMC or crystalline cellulose to any significant extent; however, this enzyme still possesses a capacity for degrading cellulose substrates in a very accessible form by removing cellobiose residues successively from the chain ends. The cellobiohydrolase was also separated from *Fusarium solani* by Wood and McCrae[120)].

Halliwell and Griffin[24, 25)] also reported that C_1 acts as a β-1,4-glucan cellobiohydrolase. They isolated a C_1 component from *T. koningii* cellulase which was free from the CMCase activity but could still solubilize cotton to an extent of 20% in 7 days when acting alone. However, when this C_1 component was mixed with a cellobiase component which had no activity on cotton, the degree of solubilization was increased to 35%. Extending the period of incubation to 21 days resulted in a marked increase in the solubilization (to 70%) when the cellobiase was contained, but no increase in the solubilization activity when only C_1 was contained.

The conclusion was that the cellobiose liberated by the C_1 action acts as an inhibitor.

Using the combined procedure of gel filtration, ion exchange, and gel electrophoresis, Pettersson et al.[73] recently separated C_1-type components from *T. viride*. They also found that the C_1 component is a hydrolytic enzyme and it removes cellobiose units according to an exo-wise mechanism. They concluded that the C_1 component is a β-1,4-glucan cellobiohydrolase. Later Berghem and Pettersson[5] and Berghem et al.[6] further purified this component and charakterized the chemical and physico-chemical properties of these components. Emert et al.[10] also reported the separation of three forms of cellobiohydrolase and one cellobiase and proposed that the function of the cellobiohydrolases is cleavage of cellooligosaccharides.

The evidence cited by the various authors to support the claim that C_1 is a cellobiohydrolase appears to be convincing. However, there remain striking differences in the extent to which highly ordered celluloses are hydrolyzed by the various C_1-type components, even when the enzymes are almost free from a contaminating CMCase activity.

3.3 C_x Components

The C_x complex of enzymes consists of the hydrolytic enzymes, which hydrolyze β-1,4-bonds in a cellulose molecule. The designation covers the random-acting endo-β-1,4-glucanase and endwise acting exo-β-1,4-glucanase. With the assistance of C_1, the β-1,4-glucanases hydrolyze crystalline cellulose; in the absence of C_1, they hydrolyze only "noncrystalline" cellulose, such as those produced by swelling, grinding, or reprecipitation from a solution[83]. The activity of the β-1,4-glucanases is usually measured by action on soluble cellulose derivatives, e.g., carboxymethylcellulose (CMC). The subscript x in C_x emphasizes the multicomponent nature of the fraction containing C_x enzymes. The β-1,4-glucanases (C_x) clearly exist as two types: (a) endo-β-1,4-glucanase with action of a random nature, the terminal linkages generally being less susceptible to hydrolysis than the internal linkages, and (b) exo-β-1,4-glucanase, successively removing single glucose units from the nonreducing end of the cellulose chain. Enzyme commission numbers have been assigned to some of the cellulase components; endo-β-1,4-glucanase is E.C. 3.2.1.4, and β-glucosidase is E.C. 2.2.1.21[36]. The latter designation does not differentiate between arylglucosidase and dimerase (cellobiase). The exo-β-1,4-glucanase has sometimes been placed in E.C. 2.2.1.21, but a new number should be assigned to it[36].

3.3.1 Endo-β-1,4-Glucanase

Many researchers carried out experiments to determine if the glucose chains of cellulose are degraded by cellulase according to an endo-wise or an exo-wise mechanism. Karrer et al.[32] and Nisizawa and Kobayashi[48,49] initially theorized that cellulase might hydrolyze the long chains of cellulose molecules according to the exo-wise mechanism. Later, however, evidence was obtained which indicates that both mechanisms are active[18]. Although Storvick and King[92] later proposed an exo-wise mechanism for hydrolysis by the cellulases of *Cellvibrio gilvus*, other works, those of Kooiman et al.[37], Reese[76], and Hash and King[28] suggested that the attack by several cellulases on polymeric substrates is basically random.

Cellulases of the following microorganisms act randomly: *Myrothecium verrucaria*[105,106,107,108,110], *Irpex lacteus*[30,31,50], *Chaetomium globosum*[103,104], *Polyporus versicolor*[68,70], *Sporotrichum pulverulentum*[2,12,13], *Cellvibrio gilvus*[91], *Pseudomonas fluorescens*[47,94], *Penicillium notatum*[67,69,71], *Fusarium solani*[114,115], *Trichoderma koningii*[23,115]. Cellulolytic enzymes of endo-β-1,4-glucanase from the culture filtrate of *Trichoderma viride* have been isolated and investigated extensively [40,42,54,56–60,62,63,65,89,97].

The proposed role of endo-glucanases in the overall process of enzymatic cellulose degradation is that they split the cellulose chains in less-ordered regions of the cellulose fiber, thus creating starting points for attack by exo-glucanases[7,118]. However, the exact role of the endo-glucanases in the degradation of cellulose fibers is not yet fully understood.

Reese and Mandels[83] have summarized the mode of action of endo-glycanases as follows: The endo-enzymes are the characteristically random-acting glucanases responsible for hydrolysis of the high-molecular-weight glucans. In such material, there are too few ends to obtain appreciable rates of hydrolysis by exo-enzymes. These enzymes act preferentially on longer chains. Acting on the soluble cellulose derivatives, these random-acting enzymes cause a rapid decrease in visçosity, with a relatively slow increase in reducing end groups. The action is not entirely random because both end linkages are somewhat less affected than the internal linkages. These endo-glucanases can also hydrolyze the soluble cellodextrins. The rate of hydrolysis increases with the degree of polymerization within the limits of substrate solubility. Cellobiose is hydrolyzed by most endo-glucanases, but not by all. The major products of endo-glucanase action are cellobiose and cellotriose.

3.3.2 Carboxymethylcellulase (CMCase)

Carboxymethylcellulase is a class of C_x which acts on the degradation of soluble cellulose derivatives, especially CMC. It is called "CMCase" for convenience by several workers[52]. The effect of chemical modification of cellulose on its susceptibility to endo-glucanase action is complex. Any action taken to increase the solubility of a polymer in water or to open the crystalline structure will increase the rate of hydrolysis. At the same time the presence of a substituent effectively prevents hydrolysis of linkages adjacent to it. In other words, the extent of hydrolysis is reduced as the number of substituents increases[46].

The enzymatic hydrolysis of CMC usually results in a loss of viscosity and an increase in the reducing power of CMC. When the rate of the viscosity reduction is relatively larger than the rate of the reducing power increase, a "more" random hydrolysis is expected to predominate. In the reverse case, a "less" or a "more exo-wise" hydrolysis is expected to predominate. Based on this observation, Gilligan and Reese[19] identified cellulase components with different degrees of randomness in the hydrolysis of CMC. They showed that these cellulase components coexist even in a single microorganism. Later, similar observations were made on several microorganisms, such as *Irpex lacteus*, *Trichoderma viride*, and *Pseudomonas fluorescens*, by Nisizawa et al.[47,50] and Okada et al.[65,66]. These workers have separated several cellulase components which differ in the rates of CMC liquefying and saccharifying activities. When increases in specific viscosities are plotted against reducing power, the slopes of the resulting curves are different for each component. Components whose curves are similar to those obtained from an acid hydrolysis, are considered "more" randomly acting. These results suggest that there are several cellulase components in each organism and that they are different in their detailed or fine specificity toward CMC[97)].

3.3.3 Exo-β-1,4-Glucan Glucohydrolase

Whitaker[107)] has reported that swollen cellulose is hydrolyzed by *Myrothecium* cellulase by an exo-wise-mechanism. His conclusion has been drawn from loss in weight of the substrate, decrease in the number-average degree of polymerization, and decrease in the weight-average degree of polymerization during incubation. However, many investigators have suggested that the attack by several cellulases on a polymeric substrate is basically random. The work of Storvick et al.[91)] has demonstrated that an exo-wise cellulase component may exist in the culture filtrate of *Cellvibrio gilvus*. Li et al.[40)], King[33)], and King and Vessal[36)] have obtained a similar component from *T. viride* cellulase. Since this enzyme component degrades cellodextrins and CMC in an exo-wise fashion, it has been called exo-glucanase. This enzyme however, attacks insoluble cellulose with difficulty, and it is presumed to be a type of CMCase.

According to Reese[78, 79)] and Barras et al.[4)], many exo-β-1,4-glucanases exist in various microorganisms in nature. These enzymes have been investigated less frequently than endo-glucanases. In general, exo-β-1,4-glucanases successively remove glucose units from nonreducing ends of the glucan. They act most rapidly on the soluble chains of four to seven units produced by the action of an endo-enzyme and, as such, play an important part in the overall digastion of cellulose to glucose. Because of their high specificity, components of enzymes have very limited action on cellulose derivatives and on glucans having mixed linkages or branches[47)]. Although they can remove one or two unsubstituted terminal units, progress down the chain is halted by any deviation from the regular structure.

3.4 β-Glucosidase

King and Vessal[36)] have summarized the mode of action of the β-glucosidases. They vary in their specificities, and the ones that act primarily on aryl-β-glucosides are not involved in cellulase action. The β-glucosidases in cellulose breakdown are those that hydrolyze cellobiose and other very short-chain β-1,4-oligoglucosides to glucose, especially those highly active on the β-dimers of glucose including cellobiose. Cellobiase is the most descriptive designation for the cellulase complex system, but a more accurate name might be "β-glucimerase", indicating the ability to act on all of the β-dimers of glucose. These enzymes do not act on insoluble cellulose. Reese et al.[82)] have compared the mode of action of β-glucosidase with that of the exo-α-glucanase. According to them, β-glucosidase usually acts on substrates ranging from cellobiose to cellohexaose. Both exo-glucanases and β-glucosidase act on cellobiose, cellotriose, and cellotetraose. The exo-enzyme acts preferentially on longer oligomers and acts by inversion. β-Glucosidase hydrolyzes dimers and trimers very rapidly, but the rate decreases markedly as the degree of polymerization increases. The β-configuration is retained by the product. Furthermore, β-glucosidase is strongly inhibited by gluconolactone, and it hydrolyzes $(1\rightarrow1)$-β, $(1\rightarrow2)$-β, and $(1\rightarrow6)$-β as well as $(1\rightarrow4)$-β linkages. Exo-glucanases have relatively high linkage specificity. The enzyme commission number assigned to the β-glucosidase is EC 3.2.1.21. But the designation does not differentiate between arylglucosidase and dimerase (cellobiase)[36].

Recently Gong et al.[20] separated three distinct cellobiase components from a commercial *T. viride* cellulase preparation by repeated chromatography on DEAE cellulose eluting with a salt gradient. They evaluated the physical properties, the kinetics, and the mechanism of these components and reported that these components are subject to product inhibition, and they hydrolyze cellobiose by a noncompetitive mechanism.

3.5 Synergism among Components

It is now well established that cellulase is a multicomponent enzyme complex, and crystalline cellulose is hydrolyzed by synergism of these cellulase components (refer to Table 1). The synergism has been discussed by many investigators. In this section, we shall focus on synergism of C_1 component (cellobiohydrolase) with other components.

Wood[116,117] separated C_1 and C_x components from *T. koningii* cellulase on DEAE-Sephadex. When each component was diluted to make it equivalent to the starting material, the fraction containing the C_x activity retained only 3% of the cellulase activity of the original culture filtrate, and C_1 component only 4%. However, when the two fractions were recombined in their original proportions, 96% of the cellulase activity of the original culture filtrate was recovered. Clearly, the activity of the cellulase complex was dependent on the synergistic action of the C_1 component with other components (see Table 2).

Halliwell and Griffin[24] fractionated and studied the synergism of enzyme fractions from the same microorganism. There was limited synergism between the C_1 component and other components of the cellulase complex, but when all the components in the complex were recombined in their original proportions, most of the activity of the original culture filtrate was recovered.

Suga et al.[93] have performed a theoretical analysis of the degradation of polysaccharides by endo- and exo-enzymes. They have considered the degradation of polysaccharide chains by endo-enzymes alone and by various combinations of endo- and exo-enzymes. The latter case is of particular interest to those who are dealing with cultures of *T. viride*, which have different exo- and endo-enzymes.

Table 2. Relative cellulase activities of the components of *T. koningii* cellulase[117]

Enzyme	Recovery of cellulase activity, %
C_x + β-glucosidase	3
C_1	4
C_1 + C_x + β-glucosidase	96
Original culture filtrate	100

The derivation of the model of Suga et al.[93)] is outlined below. The reaction between the enzyme and substrate is assumed to be of the common Michaelis-Menten type.

$$S + E \underset{k_2}{\overset{k_1}{\rightleftarrows}} ES \overset{k_3}{\longrightarrow} P + E\,. \tag{1}$$

This gives rise to the following kinetic equation;

$$\frac{dC_i}{dt} = -k_1 C_E(i-1)\,C_i + k_2 (C_{ES})_i + 2k_3 \sum_{j=i+1}^{\infty} \{(C_{ES})_j/(j-1)\}\,. \tag{2}$$

For the enzyme-glucosidic bond complex, we have

$$\frac{d(C_{ES})_i}{dt} = k_1 C_E(i-1)\,C_i - (k_2 + k_3)\,(C_{ES})_i \tag{3}$$

and for the free enzyme, we have

$$C_E = C_{Et} - \sum_{i=2}^{\infty} (C_{ES})_i\,. \tag{4}$$

Imposing the quasi steady-state approximation, we obtain from Eq. (3)

$$k_1 \left\{ C_{Et} - \sum_{i=2}^{\infty} (C_{ES})_i \right\} (i-1)\,C_i - (k_2 + k_3)\,(C_{ES})_i = 0\,. \tag{5}$$

By making use of Eqs. (3), (4) and (5), Eq. (2) becomes

$$\frac{dC_i}{dt} = -\frac{k_3 C_{Et}(i-1)\,C_i}{K_m + \sum_{j=2}^{\infty}(j-1)\,C_j} + 2\,\frac{k_3 C_{Et} \sum_{j=i+1}^{\infty} C_j}{K_m + \sum_{j=2}^{\infty}(j-1)\,C_j}\,. \tag{6}$$

When only the exo-enzyme is present, the rate of change of C_1 is given by

$$\frac{dC_1}{dt} = -\frac{k_3' C_{Et}' C_i}{K_m' + \sum_{j=2}^{\infty} C_j} + \frac{k_3' C_{Et}' C_{i+1}}{K_m' + \sum_{j=2}^{\infty} C_j}\,. \tag{7}$$

In the case of the degradation of a cellulose fragment of chain length i by various combinations of endo- and exo-enzymes, we have

$$\frac{dC_i}{dt} = -\frac{k_3 C_{Et}(i-1)\,C_i}{K_m + \sum_{j=2}^{\infty}(j-1)\,C_j} + 2\,\frac{k_3 C_{Et} \sum_{j=i+1}^{\infty} C_j}{K_m + \sum_{j=2}^{\infty}(j-1)\,C_j}$$

$$-\frac{k_3' C_{Et}' C_i}{K_m' + \sum_{j=2}^{\infty} C_j} + \frac{k_3' C_{Et}' C_{i+1}}{K_m' + \sum_{j=2}^{\infty} C_j}\,. \tag{8}$$

From the solution of these kinetic equations, the degradation index (D.I.) can be defined as

$$\text{D.I.} \equiv \frac{\mu_n(0)}{\mu_n(t)} - 1\,, \tag{9}$$

where $\mu_n(0)$ and $\mu_n(t)$ are the number average molecular weights at time = 0 and t, respectively. Figure 2 shows the numerical solution for the case of $K_m = 0.925 \times 10^{-6}$ mole of glucosidic bond ml^{-1} and $S = 10^{-3}$ g ml^{-1}. Note that the exo-enzyme by itself gives a lower rate because of the limited number of substrate sites available to it, even though its concentration is 10 times higher than that of the endo-enzyme. Again, the overall rate of action of the combined enzymes is higher than the sum of the rates of the individual enzymes.

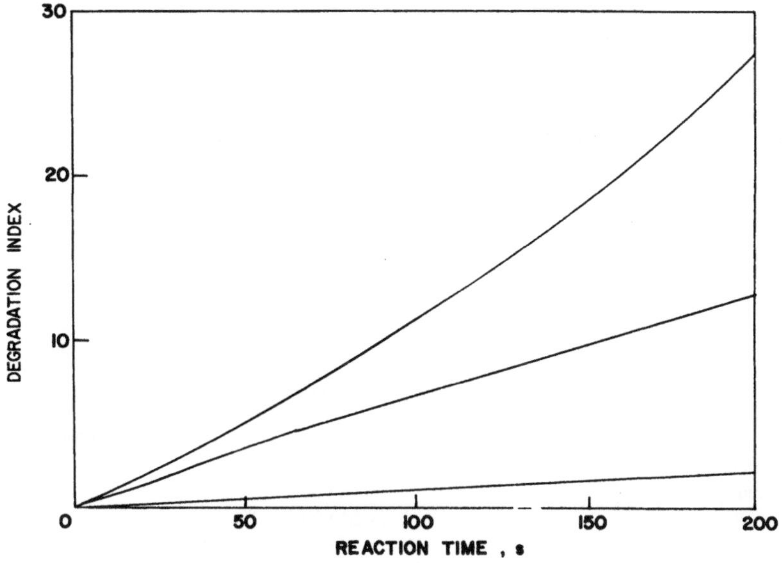

Fig. 2 Rate of change of the degradation index for exo-enzyme only (lower curve), endo-enzyme only (middle curve), and a combination of both enzymes at these concentrations. Degradation index is given in Eq. (9)[93]

The model of Suga et al.[93] does not consider branch points of linkages, inhibition by products, or diffusion and adsorption effects between the enzymes and substrates. However, the effect of the synergistic action of endo- and exo-enzymes can be evaluated on the basis of their model.

4 Physical Properties of Cellulase

4.1 Molecular Weights of Cellulase Components

Molecular weights of the various enzyme components are summarized in Table 3. The molecular weight of C_1 from *Trichoderma sp.* lies in the range of 53,000 to 62,000. The endo-β-1,4-glucanases (including C_x) show considerable variation in size, with molecular weights ranging from 5,300 to 55,000. It is possible that the smaller particles represent subunits of a normal enzyme. Except for the smaller particles, the molecular weights of most of the enzymes range from 40,000 to 50,000. Molecular weights of 50,000 to 76,000 have been reported for exo-glucanase and β-glucosidase. Ahlgren and Eriksson[1] have suggested that molecules of the glucosidase are generally larger than those of endo-glucanase. Very few estimates of the size of β-glucosidase molecules have been made. Wood[113] has obtained a molecular weight of 40,000 for the *F. solani* cellulase. The molecular weight of β-1,4-glucan cellobiohydrolase isolated from culture filtrate of *T. viride* was measured at approximately 42,000 [6].

Table 3. Estimated molecular weights of cellulase components of selected microorganisms

Microorganism	C_1 component		C_x component		β-Glucosidase	Ref.
	C_1	CBH or Avicelase	Endo-glucanase	Exo-glucanase		
Trichoderma	60,000		52,000	76,000		40)
viride	57,000		13,000			89)
	53,000		44,000			65)
			45,000			63)
		46,000				5)
		42,000				6)
			12,500			7)
			~50,000			
		48,400				21)
					76,000	20)
Trichoderma	62,000		13,000			112)
koningii			26,000			89)
			~51,000			
Fusarium solani	45,000		37,000	40,000		113)
					40,000	115)
Penicillium			35,000			1)
notatum			34,500			69)
			~35,500			
Penicillium funiculosum	57,000					86)
Sporotrichum			28,300	48,600		12, 13)
pulverulentum			~37,500			
Myrothecium			5,300			88)
verrucaria			30,000			
			55,000			
			49,000			109)

4.2 Diffusion Coefficient and Sedimentation Constant of Cellulase

Whitaker and his co-workers[109)] first attempted to determine the diffusion coefficient and sedimentation constant of a cellulase isolated from culture filtrates of the mold *Myrothecium verrucaria*. The following properties of the *Myrothecium* cellulase were determined; diffusion coefficient, $D_{W,20} = (5.6 \pm 0.1) \times 10^{-7}$; sedimentation constant, $S_{W,20} = (3.7 \pm 0.05) \times 10^{-13}$; the fractional coefficient, f/f_0, which is the frictional factor for the hydrodynamically equivalent sphere, 1.44 ± 0.02.

4.3 Size and Shape of Cellulase Molecules

The most complete physico-chemical characterization of a cellulolytic enzyme molecule was carried out by Whitaker et al.[109)] on the cellulase of *Myrothecium verrucaria*. Their data indicate that this particular enzymes is an asymmetric globular protein with a molecular weight of about 63,000. The molecule is proposed to be cigar-shaped, roughly 200 Å long and 33 Å wide at the widest point. The validity of their conclusion about the shape of this cellulase molecule is supported by comparison of hydrodynamic parameters and estimates of the size and shape derived from available viscosity and diffusion data for a well-characterized globular protein, serum albumin, and trytomysin, which is an elongated rod-shaped protein.

Cowling[8)] has estimated the sizes and dimensions of the cellulolytic enzymes of various microorganism (Table 4). In some cases the dimensions given have been calculated by the procedure of Laurent and Killander[39)] from the available gel filtration data on enzymes. In other cases the dimensions have been obtained by interpolating a log-log plot of molecular radius against molecular weight of the

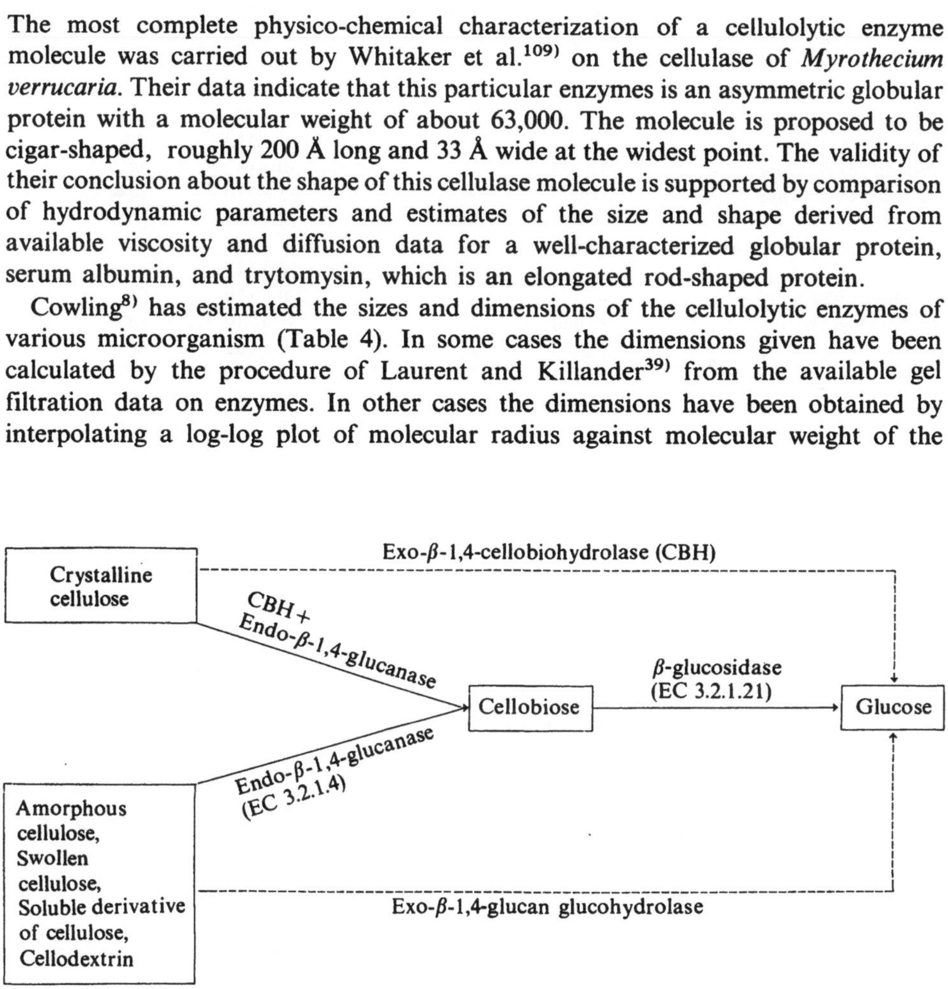

Fig. 3 Schematic summary of the pathways of cellulose degradation

⟶ Main path
--→ Side path

Table 4. Estimated size of cellulase calculated from available gel filtration (GF) and molecular weight (MW) data [8]

Organism	M. W.	Equivalent		Method of estimation
		Sphere Diameter (Å)	Ellipsoid $W \times L$ (Å)	
Aspergillus niger	—	58	32 × 192	GF
Chrysosporium lignorum	—	37	20 × 120	GF
		38	15 × 90	GF
Fomes annosus	—	42	23 × 140	GF
		38	21 × 125	GF
Myrothecium verrucaria	63,000	77	42 × 252	MW
Myrothecium verrucaria	49,000	63	35 × 208	MW
Myrothecium verrucaria	55,000	68	37 × 224	MW
	30,000	51	28 × 168	MW
	30,000	56	31 × 186	GF
	5,300	24	13 × 79	MW
Penicillium notatum	35,000	64	35 × 210	GF
Penicillium notatum	35,000	55	30 × 182	MW
Polyporus versicolor	51,000	64	35 × 210	MW
	11,400	33	18 × 108	MW
Stereum sanguinolentum	20,500	62	34 × 204	GF
	—	43	24 × 142	MW
Stereum sanguinolentum	—	37	20 × 120	GF
Trichoderma koningii	50,000	64	35 × 210	MW
	26,000	46	25 × 152	MW
Trichoderma viride	76,000	76	42 × 250	MW
	49,000	63	35 × 208	MW
Trichoderma viride	61,000	70	38 × 230	MW
Trichoderma viride				
(Rohm and Hass	—	30	17 × 100	GF
Celulase 36)	—	26	14 × 85	GF
Average for all enzymes taken together		59	33 × 200	

cellulase of *Myrothecium verrucaria* and other well-characterized globular proteins using Tanford's data[95]. Table 4 shows that dimensions of the different cellulase preparations vary considerably. If the enzymes are spherical, they range from about 25 to 80 Å in diameter with an average of 60 Å. If the enzymes are ellipsoids with an axial ratio of about 6, they range from about 15 to 40 Å in width and from 80 to 250 Å in length, giving rise to an average size of 35 × 200 Å.

5 Acknowledgement

This is Department of Chemical Engineering Contributior No. *79-192-j*, Agricultural Experiment Station, Kansas State University. This work was also supported by Department of Energy grant No. DE-FG02-79ET-00080370.

6 References

1. Ahlgren, E., Ericksson, K. E.: Acta Chem. Scand. *21*, 1193 (1969)
2. Almin, K. E., Ericksson, K. E., Pettersson, B.: Eur. J. Biochem. *51*, 207 (1975)
3. Bailey, M., Enari, T. M., Linko, M. (Eds.): Proc. Symp. Enz. Hydrol. Cellulose. Helsinki: SITRA 1975
4. Barras, D. R., Moore, A. E., Stone, B. A.: Adv. Chem. Ser. (ACS), *95*, 105 (1969)
5. Berghem, L. E. R., Pettersson, L. G.: Eur. J. Biochem. *27*, 21 (1973)
6. Bergham, L. E. R., Pettersson, L. G., Axiö-Fredricksson, U. B.: Eur. J. Biochem. *53*, 55 (1975)
7. Berghem, L. E. R., Pettersson, L. G., Axiö-Fredricksson, U. B.: Eur. J. Biochem. *61*, 621 (1976)
8. Cowling, E. B.: In: Biotech. Bioeng. Symp. No. 5, Wilke, C. R. (Ed.). p. 163. New York: Interscience 1975
9. Dellweg, H. (Ed.): Proc. V. Int. Ferment. Symp., Berlin 1976
10. Emert, G. H. et al.: Adv. Chem. Ser. (ACS) *136*, 79 (1974)
11. Eriksson, K. E.: In: Proc. Symp. Enz. Hydrol. Cellulose. Bailey, M., Enari, T. M., Linko, M. (Eds.). p. 263. Helsinki: SITRA 1975
12. Eriksson, K. E., Pettersson, B.: Eur. J. Biochem. *51*, 193 (1975a)
13. Eriksson, K. E., Pettersson, B.: Eur. J. Biochem. *51*, 213 (1975b)
14. Eriksson, K. E., Rzedowski, W.: Arch. Biochem. Biophys. *129*, 689 (1969)
15. Fan, L. T., Lee, Y. H., Beardmore, D. H.: In: Adv. Biochem. Eng. *14*, Fiechter A (Ed.), Berlin: Springer 1980, p. 101
16. Gaden, Jr., E. L. et al. (Eds.): Biotech. Bioeng. Symp. No. 5. New York: Interscience 1976
17. Gascoigne, J. A., Gascoigne, M. M.: Biological Degradation of Cellulose. London: Butterworths 1960
18. Gascoigne, J. A., Gascoigne, M. M.: In: Biological Degradation of Cellulose. p. 204. London: Butterworths 1960
19. Gilligan, W., Reese, E. T.: Canad. J. Microbiol. *1*, 90 (1954)
20. Gong, C. S., Ladisch, M. R., Tsao, G. T.: Biotech. Bioeng. *19*, 959 (1977)
21. Gum, Jr., E. K., Brown, Jr. R. D.: Biochem. Biophys. Acta, *446*, 371 (1976)
22. Hajny, G. J., Reese, E. T. (Eds.): Adv. Chem. Ser. (ACS) *95*, Washington, D.C. ACS, 1969
23. Halliwell, G.: Biochem. J. *95*, 270 (1965)
24. Halliwell, G., Griffin, M.: Biochem. J. *135*, 587 (1973)
25. Halliwell, G., Griffin, M.: Biochem. Soc. Trans. *2*, 497 (1974)
26. Halliwell, G., Riaz, M.: Biochem. J. *116*, 35 (1970)
27. Halliwell, G., Riaz, M.: Arch. Mikrobiol. *78*, 295 (1971)
28. Hash, J. H., King, K. W.: J. Biol. Chem. *232*, 381 (1958)
29. Iwasaki, Y. Hayashi, K., Funatsu, M.: J. Biochem. *55*, 209 (1964)
30. Kanda, T., Wakabayashi, K., Nisizawa, K.: J. Biochem. *79*, 977 (1976a)
31. Kanda, T., Wakabayashi, K., Nisizawa, K.: J. Biochem. *79*, 997 (1976b)
32. Karrer, P., Schubert, P., Wehri, W.: Helv. Chim. Acta, *8*, 797 (1925)
33. King, K. W.: In: Advances in Enzymic Hydrolysis of Cellulose and Related Materials. Reese, E. T. (Ed.), p. 159. New York: Pergamon 1963
34. King, K. W.: J. Ferment. Technol. *43*, 79 (1965)
35. King, K. W.: Biochem. Biophys. Res. Commun. *24*, 295 (1966)
36. King, K. W., Vessal, M. I.: Adv. Chem. Ser. (ACS) *95*, 7 (1969)
37. Kooiman, P., Roelofsen. P. A. Sweeris, S.: Enzymologia, *16*, 237 (1953)
38. Krässig, J., Kitchen, W.: J. Polymer Sci. *51*, 123 (1961)
39. Laurent, T. C., Killander, J.: J. Chromatography, *14*, 317 (1964)
40. Li, L. H., Flora, R. M., King, K. W.: Arch. Biochem. Biophys. *111*, 439 (1965)
41. Liu, T. H., King, K. W.: Arch. Biochem. Biophys. *120*, 462 (1967)
42. Mandels, M., Reese, E. T.: Develop. Indust. Microbiol. *5*, 5 (1964)
43. Marsh, C. A.: Biochem. Biophys. Acta, *122*, 367 (1966)
44. Marsh, P. B.: Text. Res. J. *27*, 913 (1957)

45. Marsh, P. B., Merola, G. V., Simpson, M. E.: Text. Res. J. *23*, 831 (1953)
46. Nisizawa, K.: J. Ferment. Technol. *51*, 267 (1973)
47. Nisizawa, K., Hashimoto, Y., Shibata, T.: In: Advances in Enzymic Hydrolysis of Cellulose and Related Materials. Reese, E. T. (Ed.). New York: Pergamon 1963, p. 171
48. Nisizawa, K., Kobayashi, T.: J. Agr. Chem. Soc. Japan, *27*, 239 (1953a)
49. Nisizawa, K., Kobayashi, T.: J. Agr. Chem. Soc. Japan, *27*, 241 (1953b)
50. Nisizawa, K. et al.: Arch. Biochem. Biophys. *96*, 152 (1962)
51. Nisizawa, T., Suzuki, H., Nisizawa, K.: J. Ferment. Technol. *44*, 659 (1966)
52. Nisizawa, K. et al.: In: Proc. IV Int. Ferment. Symp. Terui, G. (Ed.). Osaka 1972, p. 719
53. Niwa, T., Kawamura, K., Nisizawa, K.: J. Ferment. Technol. *43*, 286 (1965)
54. Niwa, T. et al.: J. Ferment. Technol. *42*, 124 (1964)
55. Norkrans, B.: Ann. Rev. Phytopathology, *1*, 325 (1963)
56. Ogawa, K., Toyama, N.: J. Ferment. Technol. *42*, 199 (1964)
57. Ogawa, K., Toyama, N.: J. Ferment. Technol. *44*, 741 (1966)
58. Ogawa, K., Toyama, N.: J. Ferment. Technol. *45*, 671 (1967)
59. Ogawa, K., Toyama, N.: J. Ferment. Technol. *46*, 367 (1968)
60. Ogawa, K., Toyama, N.: J. Ferment. Technol. *50*, 236 (1972)
61. Ogiwara, Y., Arai, K.: J. Japan. Tech. Assoc. Pulp. Paper Ind. *21*, 209 (1967)
62. Okada, G.: J. Biochem. *77*, 33 (1975)
63. Okada, G.: J. Biochem. *80*, 913 (1976)
64. Okada, G., Nisizawa, K.: J. Biochem. *78*, 297 (1975)
65. Okada, G., Nisizawa, K., Suzuki, H.: J. Biochem. *63*, 591 (1968)
66. Okada, G. et al.: J. Ferment. Technol. *44*, 682 (1966)
67. Pettersson, G.: Arch. Biochem. Biophys. *123*, 307 (1968)
68. Pettersson, G., Cowling, E. B., Porath, J.: Biochim. Biophys. Acta, *67*, 1 (1963)
69. Pettersson, G., Eaker, D. L.: Arch. Biochem. Biophys. *124*, 154 (1968)
70. Pettersson, G., Porath, J.: Biochim. Biophys. Acta, *67*, 9 (1963)
71. Pettersson, G., Porath, J.: Methods in Enzymol. *8*, 603 (1966)
72. Pettersson, L. G.: In: Proc. Symp. Enz. Hydrol. Cellulose. Bailey, M., Enari, T. M., Linko, M. (Eds.), p. 255. Helsinki: SITRA 1975
73. Pettersson, L. G., Axiö-Fredriksson, U. B., Berghem, L. E. R.: In: Proc. IV Int. Ferment. Symp. Terui, G. (Ed.), p. 723. Osaka 1972
74. Rautela, G. S., King, K. W.: Arch. Biochem. Biophys. *123*, 589 (1968)
75. Ray, D. L. (Ed.): Marine Boring and Fouling Organisms. Seattle: University of Washington Press 1959
76. Reese, E. T.: Appl. Microbiol. *4*, 39 (1956)
77. Reese, E. T. (Ed.): Advances in Enzymic Hydrolysis of Cellulose and Related Materials. New York: Pergamon 1963
78. Reese, E. T.: J. Ferment. Technol. *43*, 62 (1965)
79. Reese, E. T.: Adv. Chem. Ser. (ACS) *95*, 26 (1969)
80. Reese, E. T.: Biotechnol. Bioeng. Symp. No. 5, 77 (1975)
81. Reese, E. T., Gilligan, W.: Text. Res. J. *24*, 663 (1954)
82. Reese, E. T., Maguire, A. G., Parrish, F. W.: Canad. J. Biochem. *46*, 25 (1969)
83. Reese, E. T., Mandels, M.: In: Series on High Polymers, Vol. 5, p. 1079. Bikales, N. M., Segal, L. (Eds.). New York: John Wiley 1971
84. Reese, E. T., Siu, R. G. H., Levinson, H. S.: J. Bacteriol. *59*, 485 (1950)
85. Selby, K.: In: Advances in Enzymic Hydrolysis of Cellulose and Related Materials. Reese, E. T. (Ed.). New York: Pergamon 1963, p. 33
86. Selby, K.: In: Biodeterioration of Materials. Walters, A. H., Elphick, J. J. (Eds.), p. 62. Amsterdam: Elsevier 1968
87. Selby, K.: Adv. Chem. Ser. (ACS) *95*, 34 (1969)
88. Selby, K., Maitland, C. C.: Biochem. J. *94*, 578 (1965)
89. Selby, K., Maitland, C. C.: Biochem. J. *104*, 716 (1967)
90. Siu, R. G. H.: Microbial Decomposition of Cellulose. New York: Reinhold Pub. 1951
91. Storvick, W. O., Cole, F. E., King, K. W.: Biochem. *2*, 1106 (1963)
92. Storvick, W. O., King, K. W.: J. Biol. Chem. *235*, 301 (1960)
93. Suga, K., van Dedem, G., Moo-Young, M.: Biotechnol. Bioeng. *17*, 433 (1975)

94. Suzuki, H., Yamane, K., Nisizawa, K.: Adv. Chem. Ser. (ACS) *95*, 60 (1969)
95. Tanford, C.: Physical Chemistry of Macromolecules. p. 317. New York: Wiley 1961
96. Terui, G. (Ed.): Proc. IV. Int. Ferment. Symp., Osaka 1972
97. Toda, S., Suzuki, H., Nisizawa, K.: J. Ferment. Technol. *46*, 711 (1968)
98. Tomita, Y., Suzuki, H., Nisizawa, K.: J. Ferment. Technol. *46*, 701 (1968)
99. Tomita, Y., Suzuki, H., Nisizawa, K.: J. Ferment. Technol. *52*, 233 (1974)
100. Toyama, N., Ogawa, K.: J. Ferment. Technol. *44*, 741 (1966)
101. Walseth, C. S.: TAPPI, *35*, 228 (1952)
102. Walters, A. H., Elphick, J. J. (Eds.): Biodeterioration of Materials. New York: Elsevier 1968
103. Watanabe, T.: J. Ferment. Technol. *46*, 299 (1968a)
104. Watanabe, T.: J. Ferment. Technol. *46*, 303 (1968b)
105. Whitaker, D. R.: Arch. Biochem. Biophys. *43*, 253 (1953)
106. Whitaker, D. R.: Canad. J. Biochem. Physiol. *34*, 489 (1956)
107. Whitaker, D. R.: Canad. J. Biochem. Physiol. *35*, 733 (1957)
108. Whitaker, D. R.: In: Marine Boring and Fouling Organisms. Ray, D. L. (Ed.), p. 301. Seattle: University of Washington Press 1959
109. Whitaker, D. R., Colvin, J. R., Cook, W. H.: Arch. Biochem. Biophys. *49*, 257 (1954)
110. Whitaker, D. R., Hanson, K. R., Datta, P. K.: J. Biochem. Physiol. *41*, 671 (1963)
111. Wilke, C. R. (Ed.): Biotech. Bioeng. Symp. No. 5. New York: Interscience 1975
112. Wood, T. M.: Biochem. J. *109*, 217 (1968)
113. Wood, T. M.: Biochem. J. *115*, 457 (1969)
114. Wood, T. M.: Biochim. Biophys. Acta, *192*, 531 (1969)
115. Wood, T. M.: Biochem. J. *121*, 353 (1971)
116. Wood, T. M.: In: Proc. IV Int. Ferment Symp. Terui, G. (Ed.), p. 711. Osaka 1972
117. Wood, T. M.: Biotech. Bioeng. Symp. No. 5, 111 (1975)
118. Wood, T. M., McCrae, S. I.: Biochem. J. 128, 1183 (1972)
119. Wood, T. M., McCrae, S. I.: In: Proc. Symp. Enz. Hydrol. Cellulose. Bailey, M., Enari, T. M., Linko, H. (Eds.), p. 231. Finland: Aulanko 1975
120. Wood, T. M., McCrae, S. I.: Carbohydrate Research, *57*, 117 (1977)
121. Youatt, G.: Text. Res. J. *32*, 158 (1962)
122. Youatt, G., Jermyn, M. A.: In: Marine Boring and Fouling Organisms. Ray, D. L. (Ed.), p. 397. Seattle: University of Washington Press 1959

Kinetics of Hydrolysis of Insoluble Cellulose by Cellulase

Yong-Hyun Lee, L. T. Fan, Liang-Shih Fan
Department of Chemical Engineering, Manhattan, Kansas State University, Kanses 66506, U.S.A.

1 Introduction ... 132
2 Kinetic Characteristics of the Heterogeneous Cellulose-Cellulase System ... 133
2.1 Mass Transfer Limitation — Diffusion of Enzymes ... 133
2.2 Adsorption and Desorption of Enzymes ... 134
2.3 Surface Reactions, Fragmentations, and Changes of Surface Areas during Hydrolysis ... 136
2.4 Product Inhibition and its Mechanism ... 138
3 Kinetic Models of Insoluble Cellulose Hydrolysis by Cellulase ... 140
3.1 Empirical Representations of the Kinetics of Enzymatic Hydrolysis ... 140
3.2 Michaelis-Menten Mechanism Applied to an Insoluble Cellulose-Cellulase System ... 141
3.3 McLaren's Approach to Enzyme Reactions in a Structurally Restricted System ... 146
3.4 Kinetic Models Based on the Combination of the Enzyme Adsorption Equation and the Michaelis-Menten Type Equation ... 148
3.5 Kinetic Model Based on the Depolymerization Mechanism ... 154
3.6 Distributed Parameter Kinetic Models ... 155
4 Kinetics of the Reactions of Cellulases with β-1,4-Oligoglucosides ... 159
5 Kinetics of β-Glucosidase ... 162
6 Summary ... 163
7 Acknowledgement ... 164
8 Symbols ... 165
9 References ... 167

The article covers the literature on the kinetic aspects of the cellulose-cellulase system. The kinetic characteristics of this heterogeneous enzyme reaction are described first and then the kinetic expressions for the hydrolysis of insoluble cellulose by cellulase. In addition, the kinetics of the reactions of cellulases with soluble cellooligosaccharides is discussed.

1 Introduction

Enzyme kinetic studies have been concerned mainly with the homogeneous system, i.e. the action of a soluble enzyme on a soluble substrate. However, many important enzyme reactions occur in heterogeneous systems. Some of the heterogeneous enzyme reactions involve the action of insoluble enzymes (or bound enzymes) on soluble substrates and the action of soluble enzymes on insoluble substrates. Some mitochondrial enzymes, plant root surface phosphatase and immobilized enzymes are examples of the insoluble state enzymes which act on soluble substrates. Enzymes involved in the digestion of foodstuffs in animal intestine and enzymatic hydrolysis of cellulose and starch are examples of the action of the soluble enzymes on insoluble substrates. Considerable research on the action of insoluble enzymes on soluble state substrates has been performed with immobilized enzyme systems, whereas the action of a soluble enzyme on an insoluble substrate remains as one of the least known areas of enzyme kinetics[40)].

Some of the kinetic studies of cellulases have been carried out by employing cellulose modifications, e.g., carboxymethylcellulose (CMC) for an increased solubility[9)]. The results obtained from the cellulose modifications cannot be applied directly to insoluble cellulose because the physical properties of cellulose are altered significantly by the presence of substituent groups. Some investigators have conducted kinetic studies using soluble oligosaccharides, but their results provide only an approximate picture of the kinetics of the enzyme hydrolysis of a soluble cellulose[29, 33, 46, 54, 61)].

Recently, several reports dealing with the kinetics of cellulase on insoluble native cellulose have been published. Some of them are rather empirical[4, 33, 56)]. Several researchers[2, 5, 15, 19, 20)] attempted to explain the kinetic behavior of the insoluble cellulose-cellulase system by resorting to the Michaelis-Menten type kinetics involving product inhibition effects. Kinetic models, each based on a combination of enzyme adsorption and Michaelis-Menten kinetics, have also been proposed[8, 21, 22, 23, 39)]. Such models generally assume that the enzyme is adsorbed on the surface of the substrate and that the rate of digestion is proportional to the concentration of the adsorbed enzyme. Some kinetic models proposed take into account the mode of action of each cellulase component on cellulose molecules with different degrees of polymerization[31, 43, 53)]. These models are based on the Michaelis-Menten type kinetics for concurrent random and endwise attack of the substrate involving end-product inhibitions and several types of enzymes.

A limited number of distributed parameter models, which consider both the mass transfer and reaction, have been proposed[48, 52)]. However, experiments to test the validity of the proposed models were carried out with dextran-dextranase systems rather than with cellulose-cellulase systems. A feedback inhibition model of cellulose degradation by *Thermoactinomyces*, involving different enzyme fractions with an associated synergism, has been developed[3)]. This model is not elaborated here because it is similar to the microbial growth model.

The kinetics of the enzymatic hydrolysis of insoluble cellulose primarily depends on three groups of factors:

the structural features of cellulose,
the nature of the enzyme system employed and
the mode of interaction between cellulose and enzyme.
It appears that a kinetic model, which considers all these factors, has not been proposed. To derive a mechanistic kinetic model, the structural features of cellulose and the mode of action of the cellulase complex should be fully investigated. Furthermore, kinetic characteristics of the heterogeneous cellulose-cellulase system, such as mass transfer, adsorption and desorption of the enzyme, surface reaction and product inhibition, should be examined.

We have reviewed the major chemical and physical features of cellulosic materials as substrates for the enzymatic hydrolysis[10] and the properties and mode of action of cellulase[32]. This review covers the literature on the kinetic aspects of the cellulose-cellulase system. The kinetic characteristics of this heterogeneous enzyme reaction are described first and kinetic expressions for the hydrolysis of insoluble cellulose by cellulase are then reviewed. In addition, the kinetics of the reactions of cellulases with soluble cellooligosaccharides is discussed. Such information is essential in developing a kinetic model of the enzymatic hydrolysis of native insoluble cellulose.

2 Kinetic Characteristics of the Heterogeneous Cellulose-Cellulase System

2.1 Mass Transfer Limitation — Diffusion of Enzymes

The formation of an enzyme-substrate complex is a prerequisite to the enzymatic hydrolysis of native cellulose. Since cellulose is an insoluble and structurally complex substrate, the formation of an enzyme-substrate complex can be achieved only by the diffusion of enzymes into the complex structural matrix of cellulose. Because of the heterogeneous nature of the process, the overall rate of the reaction conceivably could be influenced by mass transfer resistances, such as the bulk phase resistance, the resistance through the surface film around cellulose particles, and the resistance through capillary pores of cellulose particles. The resistances through the bulk phase and film adjacent to the particles depend on the size of cellulose particles, cellulase concentration, and hydrodynamic conditions in the reactor, such as the intensity of agitation and the Reynolds number flow through the reactor.

Van Dyke[56], experimenting to determine the magnitude of the bulk phase resistance, observed that agitation intensity had little effect on the hydrolysis, provided that the cellulose particles were completely suspended. The minimum rate of agitation (RPM) to suspend 8 wt percent cellulose was 100, which corresponds to the Reynolds number of 1,200. The Reynolds number, corresponding to a transition from the laminar to the turbulent flow pattern, for a three-bladed propeller used by Van Dyke in a baffled reactor was 1,000. Other investigators[21,26] obtained similar results in their experiments studying mass transfer limitations. When agitation speed exceeded 100 RPM, mass transfer resistances were negligible in a batch reactor containing pure cellulose as a substrate if its concentration was 2–10%. This indicates that the bulk and film resistance can be made virtually non-existent under certain experimental conditions, including adequate mixing, proper particle size of cellulose, and proper enzyme concentation.

No experimental studies have been reported on the pore diffusion resistance inside cellulose particles. Kim[26] suggested that the macromolecular nature of enzyme molecules may cause pore diffusion resistance to be insignificant. In other words, the enzyme does not diffuse through the pores of cellulose particles because the size of an enzyme molecule is larger than most of the capillary pores of cellulose.

2.2 Adsorption and Desorption of Enzymes

Since adsorption of enzymes on the cellulose surface is a prerequisite for the hydrolysis of cellulose, it should be studied in detail if we wish to understand the kinetic behavior of the heterogeneous cellulose-cellulase system. One of the earliest works on the adsorption of cellulase by cellulose was carried out by Halliwell[17]. He reported that the aqueous phase becomes relatively free of enzyme even immediately after mixing cellulase and substrate because of the adsorption of the enzyme by the cellulose; the extent of adsorption depends on the initial quantity of cellulose present and on subsequent incubation of the mixture.

Mandels et al.[34] also observed that cellulose strongly adsorbs cellulase under conditions optimal for enzyme actions, and that the extent of the enzyme or protein uptake is proportional to the cellulose concentration. More than 90% of the enzyme and soluble protein were adsorbed from an unconcentrated cellulase solution by a mixture containing 10% cellulose and about 90% of the enzyme, and more than 70% of protein were adsorbed from a concentrated cellulase solution by the same mixture. The actual uptake ranged from 0.005 to 0.064 mg protein per mg of cellulose. The maximum uptake of cellulase per volume of cellulose was attained when the cellulose concentration was lowest. Adsorption increased as the average particle diameter decreased from 50 µm to 6.7 µm.

The adsorption of *Trichoderma* cellulase in aqueous solution on three kinds of cellulose was studied by Peitersen *et al.*[44]. They found that the adsorption of protein and enzyme from the solution is largely independent of pH but strongly dependent on the temperature and type of cellulose. They used the following Langmuir isotherm type equation to relate the adsorbed enzyme concentration to the initial free enzyme concentration:

$$E_{ads} = \frac{K_p E_{ads,m}}{1 + K_p E_0} E_0 \qquad (1)$$

where

E_0 = protein concentration in the supernatant
E_{ads} = adsorbed protein
$E_{ads,m}$ = maximally adsorbed protein
K_p = constant.

Peitersen et al.[44] observed that, generally, an increase in the temperature reduced the maximum adsorbed protein. At the optimal hydrolysis temperature of 50 °C, the maximally adsorbed protein by Avicel, that by Solka Floc SW40 and that by ball milled Solka Floc were 0.027, 0.019, and 0.049 mg protein per mg cellulose, respectively. They postulated that the variation in these values might stem from the differences in the total available surface areas among the various types of cellulose.

Others have observed differences in adsorbability among the cellulase components[14,63,64]. Mandels *et al.*[34] reported that component (C_1), which has an activity on filter paper, was more strongly adsorbed than component (C_x), displaying an activity on CMC. Wilke and Mitra[63] made a similar observation after determining adsorption equilibria of both C_1 and C_x activities at various concentrations of milled newsprint. The adsorption characteristics of cellulase components on bagasse, which contains appreciable amounts of cellulose and hemicellulose, were presented and discussed by Ghose and Bisaria[13] in the light of the role of the adsorption characteristics on hydrolysis. They reported that simultaneous adsorption of exo- and endo-glucanase on hydrolyzable cellulose is the causative factor of hydrolysis that immediately follows the adsorption step, supporting the postulate of synergistic enzyme action.

The time course of the adsorption of cellulase in a solution by cellulose, determined by Mandels et al.[34], is shown in Fig. 1. They mixed concentrated cellulase with 10% milled cellulose, incubated the resultant mixture in a shaker at 50 °C, and adjusted the pH to 4.4 with 0.05 M citrate buffer. Fig. 1 shows that the initial uptake of cellulase was rapid with most of the enzyme removed from the solution after 20 min, but a slow uptake continued for several hours. Maximum adsorption was reached in approximately 8 h. Once the cellulose was supersaturated with cellulase, an negligible uptake was observed; here thereafter, the bulk concentration of the cellulase in the solution remained almost constant and eventually increased slightly, indicating that the cellulase was released from the cellulose by the subsequent digestion of cellulose.

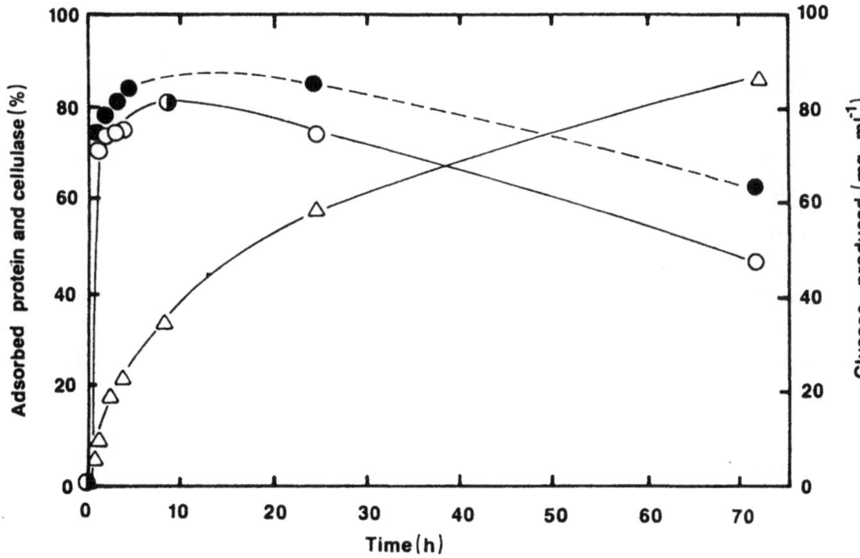

Fig. 1 Effect of time on the adsorption of cellulase by cellulose[34]

Other interesting experimental results on the behavior of adsorbed cellulase obtained by Kim[26] are shown in Fig. 2. It depicts typical digestion curves of a 6% suspension of ball milled Solka Floc under various experimental conditions. Curve (A) shows the original batch reducing sugar production, curve (B) the reducing sugar production when the supernatant was removed and a buffer added. To obtain curve (B), the reaction mixture was centrifuged after 1 h of reaction time. Then, a fresh buffer solution was added to the precipitate, and the reaction was continued in a shaker incubator. As shown in Fig. 2, the reducing sugar production rate was nearly equal to that obtained in the original batch [Curve (A)]. This finding led Kim[26] to conclude that the enzyme initially adsorbed onto cellulose is primarily

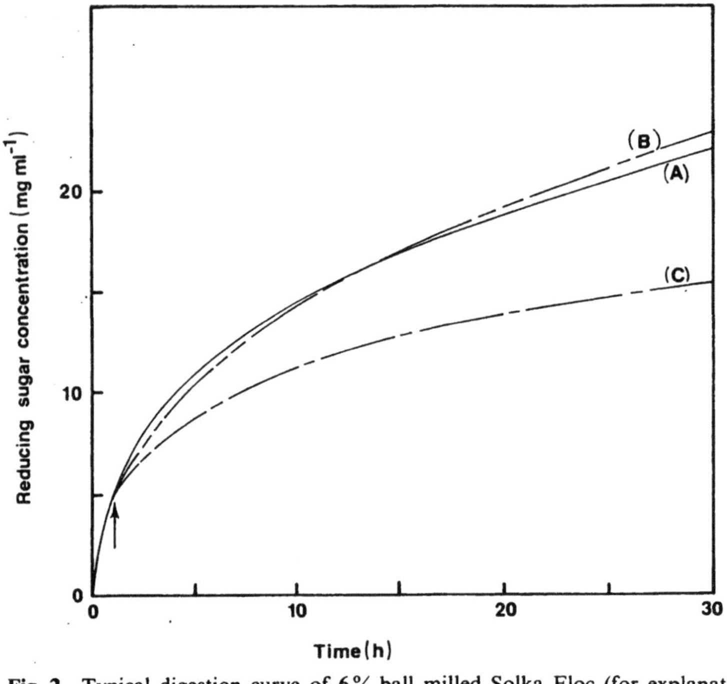

Fig. 2 Typical digestion curve of 6% ball milled Solka Floc (for explanation of curves (A), (B) and (C) see text[26])

responsible for cellulose digestion. Nevertheless, his conclusion needs further verification because when the supernatant is removed, most of the enzyme may exist in the adsorbed state. Curve (C) shows the reducing sugar production after the supernatant was removed and a buffer solution and a glucose solution with a concentration of 20 mg ml^{-1} were added. Curve (C) illustrates the effect of product inhibition on the cellulase action.

2.3 Surface Reactions, Fragmentations, and Changes of Surface Areas during Hydrolysis

The reaction between cellulose and cellulase is heterogeneous, and thus, the rate of reaction is expected to the proportional to the amount of cellulose surface accessible to the enzyme molecules. Cowling[6] and Cowling and Kirk[7] proposed that the accessible cellulose surface is the most important structural feature of cellulose for enzymatic hydrolysis. Stone et al.[50] reported that there is indeed a linear relationship between the initial rate of reaction and the initial surface area available to the enzyme molecules. Phosphoric acid swollen cellulose was used as the substrate. Meanwhile, King[27] reported a different relationship between the rate of crystalline cellulose digestion and the available surface area. He determined the size distribution after each time increment and the corresponding amount of cellulose solubilized by multiplying the volume of each particle by the change in number of such particles per unit volume of solution. The surface area of particles in each size class was then calculated assuming that the particles were spherical. He found that a linear relationship exists between the rate of degradation of crystalline hydrocellulose particles and the square of their surface areas, which suggests that the enzyme action is not exclusively a surface phenomenon.

According to King[28], the large cellulose fiber particles were fragmented into smaller ones (800–1500 Å) during the initial stage of reaction. These observations coincide with the early works of Halliwell[18] and Marsh[37], who reported that the early phase of enzymic breakdown of cellulose is characterized by the formation of a large number of very short fragments, which increases to a maximum and then gradually decreases by conversion to soluble sugars. Therefore, in addition to the initial surface area of the cellulose, other factors, such as fragmentation of a cellulose particle and the other structural features of cellulose, should also be considered in the kinetic study of the degradation of insoluble cellulose.

Fan et al.[11] measured the change in the specific surface area of Solka Floc during its hydrolysis by the filtrate of *Trichoderma* cellulase. They used a solvent drying technique[38] to measure the water-swollen surface area of cellulose. By

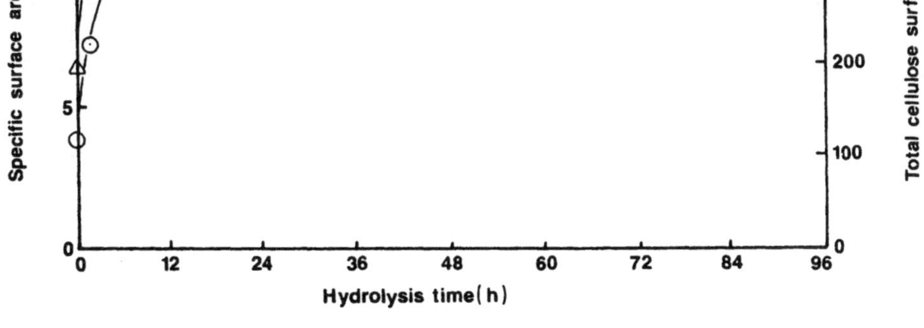

Fig. 3 Change of the specific surface area and total surface area during hydrolysis[11]

applying the BET equation to nitrogen adsorption data, they determined the specific surface areas during hydrolysis, as shown in Fig. 3. The total surface area was determined by multiplying the residual cellulose weight per liter by the specific surface area. The specific surface area drastically increased from about $4 \, m^2 \, g^{-1}$ of cellulose to $10.5 \, m^2 \, g^{-1}$ during the first 6 h. After that, it slowly increased until 24 h of hydrolysis, then the value leveled off to $12.2 \, m^2 \, g^{-1}$ of cellulose. The initial increase of the specific surface area was observed to be mainly due to the fragmentation and particle size reduction. The reaction rate slowed down even when a substantial amount of surface area existed. Fan et al.[11] consequently concluded that the surface area may not be the main limiting factor in cellulose hydrolysis as had been thought. To explain the slow-down of the reaction at the later phase, they have postulated that the surface area is composed of two fractions, active and inactive, and that the rate of hydrolysis may be dependent on the active surface area rather than on the overall surface area.

2.4 Product Inhibition and its Mechanism

In enzyme reactions, the accumulated end products usually inhibit the rate of the forward reaction. A comprehensive review[35] has been published on the inhibition of cellulases and β-glucosidases, which is caused by physical factors, chemical actions, and natural inhibitors. We review only the product inhibition in this section.

Mandels and Reese[35] reported that cellobiose inhibited the hydrolysis of cellulose by filtrates of the majority of the 36 organisms they tested. The inhibitive action of cellobiose was reported to proceed competitively; in other words, cellobiose was competing with the substrate cellulose for active sites on the enzyme. According to Mandels and Reese[35], the degree of inhibition depends on the relative affinities of the inhibitor and substrate for the active sites and on the ratio of the concentrations of the inhibitor and substrate. The inhibitory effect of products varies with the organism from which the cellulase is derived.

The product inhibition in the cellulose-cellulase system was observed to be large when the large soluble oligomer (cellopentaose) was present and to decrease rapidly with a decrease in the size of product molecules[42]. According to Ghose and Das[15], glucose mildly inhibits the hydrolysis of finely milled and heat-treated cellulose catalyzed by *Trichoderma viride* cellulases to an extent of only 40% at a concentration of 30%. On the other hand, cellobiose inhibits the reaction fairly severely to an extent of 50% even at a very low concentration. Ghose and Das[15] also investigated the cellulase inhibition mechanism. They analyzed their experimental results based on Henry's equation by assuming that the cellulase-cellulose system is an uncomplicated one-substrate reaction. They found that the plot of

$$\frac{1}{x} \log \left(\frac{a}{a - x} \right) \quad \text{against} \quad \frac{t}{x}$$

for cellulosic substrate suspensions of 2%, 5%, and 10% in concentrated *T. viride* cellulase solution did not yield the expected linearity (where x is the concentration).

of cellulose consumed, a the initial cellulose concentration, and t the reaction time). Thus, it was concluded that an inhibition mechanism is involved. Ghose and Das[15] have also concluded from analyses of the Lineweaver-Burk plot that the nature of inhibition can be represented by competitive product inhibition. The linear form of the product inhibition equation can be written as

$$\frac{1}{v} = \frac{K}{V}\left(1 + \frac{P}{K_p}\right)\frac{1}{(S)} + \frac{1}{V} \tag{2}$$

where

v = reaction rate
V = maximum reaction rate
P = product concentration
K = Michaelis-Menten constant
K_p = constant
(S) = substrate concentration .

Fig. 4 shows the reciprocal or Lineweaver-Burk plots obtained by Ghose and Das[15] for no-inhibition, 10 g l^{-1} glucose inhibition, and 5 g l^{-1} cellobiose inhibition. From these plots, they obtained: (a) $K = 3.45$ mg ml^{-1} and $V = 1.3$ for the case of no inhibition, (b) $K_{p,\,cellobiose} = 28.90$ mg ml^{-1} for the case with 5 mg ml^{-1} cellobiose inhibition, and (c) $K_{p,\,glucose} = 24.21$ mg ml^{-1} for the case of 10 mg ml^{-1} glucose inhibition.

Meanwhile, Stuck[51] and Howell and Stuck[20] suggested an enzyme inhibition mechanism which is different from that of Ghose and Das[15]. In developing a product inhibition model to describe the hydrolysis of cellulose by the *Trichoderma viride* enzyme system, they assumed that the noncompetitive inhibition by cellobiose dominates the reaction kinetics, and reported that under various experimental conditions, their noncompetitive model can predict the progress of the reaction for an extended time more adequately than a competitive product inhibition model.

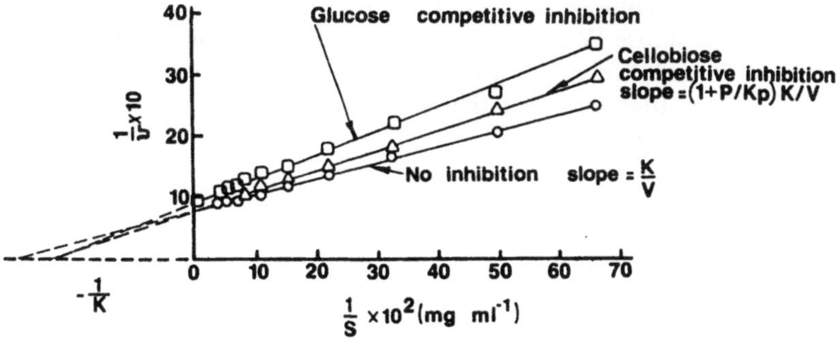

Fig. 4 Lineweaver-Burk plots for non-inhibition, glucose inhibition and cellobiose inhibition in <25 μ cellulose, enzyme activity 2.0 FP, Temp. 50 °C [15]

3 Kinetic Models of Insoluble Cellulose Hydrolysis by Cellulase

3.1 Empirical Representations of the Kinetics of Enzymatic Hydrolysis

An early empirical model was proposed by Karrer et al.[25], who used enzymes from the digestive juice of snails. They found that these enzymes hydrolyze regenerated cotton cellulose according to the Schultz equation. This equation expresses the rate of the reaction by $k/\sqrt{t}$, where k represents the rate constant. Miyamoto and Nisizawa[41] also investigated the kinetics of the hydrolysis of colloidal hydrocellulose catalyzed by cellulase preparations from several fungi. Analyzing their results according to the approach by Karrer et al.[24,25], they found that the hydrolysis proceeds according to the following equation:

$$X = kt^m e^n \,, \tag{3}$$

where
X = extent of hydrolysis after time t
k = empirical rate constant
t = hydrolysis time
e = concentration of enzyme
n, m = constants which are characteristics of each cellulase source.

Another empirical model proposed by King[27] shows that the initial rate of hydrolysis of highly crystalline cellulose can be described by the Schultz law, which states that the extent of the reaction is proportional to the square root of time, i.e.,

$$X = k\sqrt{t} \,, \tag{4}$$

where
X = extent of hydrolysis
k = rate constant
t = time.

Rautela and King[45] reported that the use of different types of crystalline cellulose results in different rate expressions. According to them, the rates of enzymatic dissolution of crystalline cellulose I and II obey the Schultz law, the rate of crystalline cellulose II/IV follows the second-order kinetics and the rate of crystalline cellulose IV follows no recognizable reaction order.

Ghose[12] and Walseth[57] gave a more general description than that given by King[27]. They theorized that the initial fast rate of hydrolysis may be represented by the empirical expression

$$X = kt^n \,, \tag{5}$$

where
X = extent of hydrolysis
t = time
k = rate constant
n = constant.

Ghose[12] explained that several factors, such as the heterogeneity of the substrate, interactions between the C_1 and C_x enzymes, and build-up of resistant cellulose during the course of hydrolysis, make it doubtful that the kinetic pattern, as suggested for the initial phase of reaction, is applicable to later stages of hydrolysis.

Kinetic expressions, which are valid over the entire reaction period, were proposed recently by Van Dyke[56] and Brandt et al.[4]. Van Dyke used the concept of multiple substrates within the matrix to formulate a model, which assumes that the formation of products is simply the summation of pseudo first-order reactions between the enzyme and substrate; that is,

$$\frac{dP}{dt} = \sum_{i=1}^{n} k_i C_i , \tag{6}$$

where

P = product concentration

C_i = concentration of cellulose component i

k_i = rate constant associated with component i.

To determine the concentration of the various components, Van Dyke divided the reaction progress curve into several regions which were considered to be constant rate regions. He then extrapolated the reaction curve in each of the divided regions to time zero. Thus, several intercepts that correspond to the different cellulose components have been obtained. He reported three cellulose components: insoluble amorphous cellulose, crystalline cellulose, and inaccessible cellulose.

The model proposed by Brandt et al.[4] is similar to Van Dyke's, in that it also consists of a set of first-order reactions to describe the process. Concentrations of different forms of cellulose were determined by location of inflection points.

The works of Van Dyke[56] and Brandt et al.[4] represent significant attempts to develop readily usable, kinetic expressions for the overall rate of cellulose hydrolysis. However, their models do not take into account various enzyme-substrate reaction mechanisms, e.g. enzyme-substrate complex formation. Thus, their models should be considered as oversimplified semiempirical models.

3.2 Michaelis-Menten Mechanism Applied to an Insoluble Cellulose-Cellulase System

Amemura and Terui[2] attempted to apply the enzyme kinetics of the Michaelis-Menten type directly to an insoluble cellulose-cellulase system, using a cellulolytic enzyme obtained from *Penicillium variable*. This enzyme contains three cellulase components and degrades native cellulose directly into glucose without accumulating cellooligosides. Therefore, this enzyme system is convenient for the kinetic study of insoluble substrates, as both the decrease in the weight of the substrate and the increase in the reducing power in the incubation mixture can be readily determined as a measure of cellulase activity[1].

The effect of substrate concentration on the reaction rate, v, was investigated by Amemura and Terui[2] either by changing the initial substrate concentration, $[S]_0$, or by following the residual substrate concentration, $[S]$. Unexpectedly their data obeyed the classical Michaelis-Menten model as shown in Fig. 5. This figure plots $1/v$ against $1/[S]$ and the data are fitted to the integrated Michaelis-Menten

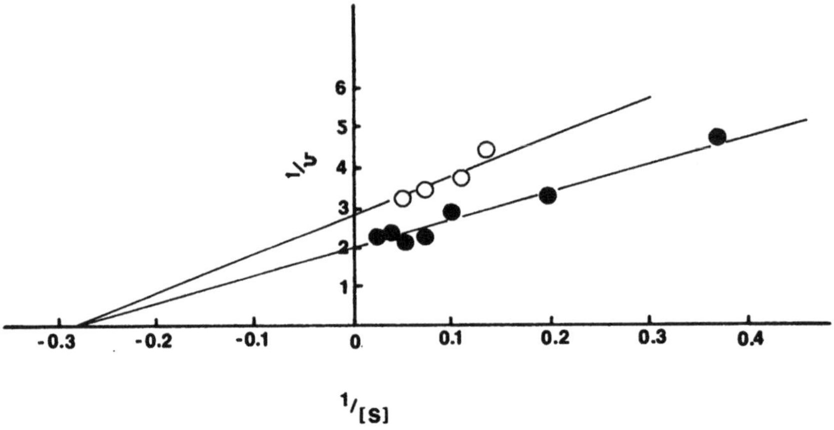

Fig. 5 Relationship between the substrate concentration and the initial rate of decomposition expressed by the Lineweaver-Burk plot[2)]

equation. On the basis of this result, Amemura and Terui[2)] stated that neither the adsorption rate nor the penetration rate of enzyme molecules limits the overall rate of substrate dissolution and that the overall number of active sites in the substrate open for the chance of enzyme attack remains nearly constant throughout the reaction. Note that their observations were based on the initial hydrolysis rate, where the structural changes can be negligible.

Since the cellulosic substrate usually used in hydrolysis contains 70% crystalline cellulose, which is difficult to be digested by cellulase, the rate equation based on the initial rate cannot be applied directly to the entire range of the reaction. Thus, Amemura and Terui[2)] introduced the concept of the effective substrate concentration, $[S]_a$, as follows:

$$[S]_a = \alpha[S]_t \,, \tag{7}$$

where

α = constant less than unity

$[S]_t$ = total substrate concentration .

Replacing $[S]_a$ in the Michaelis-Menten equation with $[S]_t$ gives rise to the following integrated rate expression.

$$t_r = \frac{1}{V'} \{[S]_i(1 - r) - K'_m \ln r\} \,, \tag{8}$$

where

$$V' = \frac{V}{\alpha}$$

$$K'_m = \frac{K_m}{\alpha}$$

$$r = \frac{[S]_{i+t_r}}{[S]_i}$$

$[S]_i$ = substrate concentration at time i

$[S]_{i+t_r}$ = substrate concentration at time $i + t_r$

t_r = time required for reducing the cellulose concentration from $[S]_i$ to $[S]_{i+t_r}$.

A value of 3.75 g l^{-1} was obtained for the modified Michaelis-Menten constant, K'_m, from the initial rate experiments.

Amemura and Terui[2] also observed inhibition by the glucose produced, which prompted them to introduce the product inhibition term in their kinetic model. Their analysis of the time course of reaction indicates that the inhibition is of a competitive nature, and that the inhibitor concentration, $[I]$, is increased as the reaction proceeds due to the accumulation of glucose, which is proportional to $[S]_0 - [S]$, i.e., $[I] = c([S]_0 - [S])$. Eventually, they derived the following integrated form of the amended Michaelis-Menten equation to fit the data.

$$t_r = \frac{1}{V'}\left\{\left(1 - \frac{cK'_m}{K'_i}\right)[S]_i(1-r) - K'_m\left(1 + \frac{c[S]_0}{K_i}\right)\ln r\right\}. \tag{9}$$

Fig. 6 shows the relationship between t_r and $[S]_i$, which yields

$K'_i = 1.50\,c$, $V' = 1.66$ g l^{-1} h^{-1}, $[S]_0 = 10.4$ g l^{-1},

$K'_m = 3.57$ g l^{-1}, $r = 0.8$.

From these results, Amemura and Terui[2] concluded that the Michaelis-Menten equation is applicable even for the degradation of an insoluble substrate by the cellulase if the total substrate concentration, $[S]_t$, is replaced by the effective substrate concentration, $[S]_a$, or if the accessibility of cellulase to the substrate is

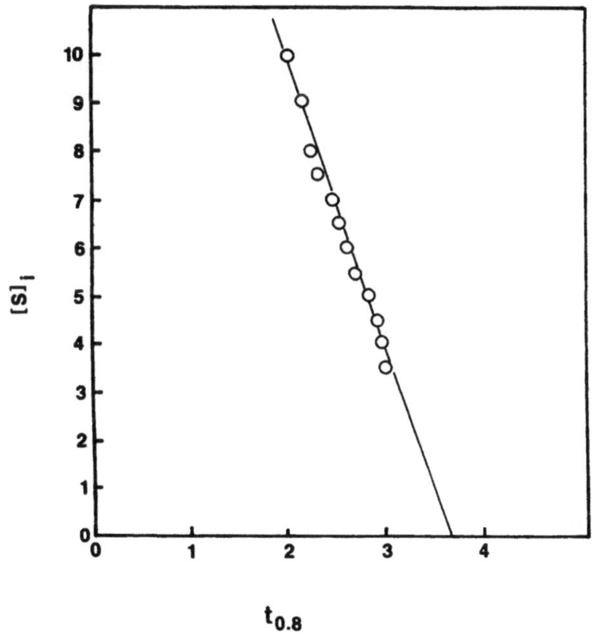

Fig. 6 Plot of Eq. 9 and experimental result (○), showing the competitive nature of product (glucose) inhibition[2]

taken into account. Their work represents an early study on the kinetics of enzymatic hydrolysis of cellulose, which attempted to apply the Michaelis-Menten kinetics with the product inhibition and substrate multiplicity. Although their model is based on the assumption of a constant α, no definitive experimental results have been obtained to determine the value of this constant.

The importance of product inhibition on the kinetics of the enzyme system derived from *Trichoderma viride* was demonstrated by Ghose and Das[15]. They determined the initial hydrolysis rate of Solka Floc in the presence and absence of glucose and cellobiose and observed a significant product inhibition. Their competitive product inhibition model has already been reviewed in a previous section.

Meanwhile, Howell and Stuck[20] developed a non-competitive inhibition model to describe the hydrolysis of cellulose by the *Trichoderma viride* enzyme system. This enzyme system was deficient in β-glucosidase, and thus, the predominating end product over short to moderate time would be cellobiose. It was assumed that non-competitive inhibition by cellobiose dominates the reaction kinetics. The model is based on the following mechanism;

$$[G_2] + E_1 \underset{k_{-1}}{\overset{k_{+1}}{\rightleftharpoons}} E_1[G_2], \tag{10}$$

$$E_1[G_2] \xrightarrow{k_2} E_1 + G_2, \tag{11}$$

$$E_1[G_2] + G_2 \underset{k'_{-3}}{\overset{k'_{+3}}{\rightleftharpoons}} E_1[G_2]G_2, \tag{12}$$

where

$[G_2]$ = cellobiose concentration in polymerized form
G_2 = cellobiose concentration
E_1 = enzyme concentration.

The application of the pseudo-steady state assumption to the enzyme complexes has yielded the following expression relating the substrate and product concentrations to time.

$$Vt = K_s\left(1 + \frac{S_0}{K_i}\right) \ln \frac{S_0}{S_0 - P} + \left(1 - \frac{K_s}{K_i}\right) P + \frac{P^2}{2K_i}, \tag{13}$$

where

S_0 = initial cellulose concentration, mol of anhydrocellobiose
P = G_2 = product (cellobiose) concentration
K_s = dissociation constant for the ES complex
K_i = dissociation constant for the EP complex
V = maximum reaction rate.

Kinetic parameters in this expression were determined by Howell and Stuck[20] from the data obtained in batch experiments with durations of up to 15 h, by plotting the data on the Foster-Nieman coordinates. The resultant values of the parameters are: $K_i = 4.68 \times 10^{-3}$ mol cellobiose l^{-1}, $K_m = 9.56 \times 10^{-2}$ mol polyanhydrocellulose l^{-1}, $V = 2.92 \times 10^{-2}$ mol cellobiose $l^{-1}\,h^{-1}$ at an enzyme concentration equivalent to a filter paper activity of 1.5 units.

As a first approximation, Howell and Struck[20] suggested the equation for reaction progress with non-competitive inhibition as

$$t = 3.27\left(1 + \frac{[G_2]_0}{4.68 \times 10^{-3}}\right) \ln \frac{[G_2]}{[G_2]_0 - G_2} - 6.58 \times 10^{-3} G_2 + \frac{[G_2]^2}{9.56 \times 10^{-3}} . \tag{14}$$

The applicability of this model is limited to the initial short periods of time up to 10 h, corresponding to the extent of hydrolysis of 50 ~ 60%. After this period the substrate multiplicity becomes significant in hydrolysis. Furthermore, the substrate concentrations in the experiments were lower than that usually employed in practice, namely, around 50 g l^{-1}.

Brown and Waliuzzaman[5] studied the kinetics of enzymatic hydrolysis of waste newspaper, employing a similar competitive product inhibition model and the Foster-Niemann plot. The values of the parameters of the model obtained were almost identical to those for pure Solka Floc cellulose reported by Howell and Stuck[20]. Brown and Waliuzzaman suggested that the data from the studies of pure cellulose would be directly applicable to the estimation of waste cellulose reactions.

Mangat[36] and Howell and Mangat[19] showed that the product inhibition kinetic model represents the experimental data well for the first 12 h of hydrolysis, but, thereafter, it overpredicts the rate of hydrolysis. To overcome this limitation, a model postulating inactivation of the adsorbed enzyme-substrate complex which blocks further hydrolysis was proposed by Howell and Mangat[19]. The main qualitative features of their model included: i) product inhibition by cellobiose is highly significant and controls the initial fast hydrolysis reaction, ii) a final portion of the substrate appears to be completely inaccessible to the enzyme, and iii) enzyme inactivation occurs according to a process other than product inhibition. Based on these premises, they proposed the following mechanism:

$$E + S \underset{k_{-1}}{\overset{k_1}{\rightleftharpoons}} ES , \tag{15}$$

$$ES \xrightarrow{k_2} P , \tag{16}$$

$$E + P \underset{k_{-3}}{\overset{k_3}{\rightleftharpoons}} EP , \tag{17}$$

$$ES \xrightarrow{k_4} D . \tag{18}$$

Eqs. (15)–(17) describe the competitive product inhibition mechanism, and Eq. (18) shows that the *E-S* complex deactivates or denatures, forming an inactivated enzyme-substrate complex (*D*). This inactivated complex plays no further part in the hydrolysis reaction. By applying a quasi steady-state assumption, the following integrated form of the equation was derived.

$$k_2 t = \frac{A}{\varepsilon} \ln\left[\frac{(E_0)}{(E_0) - \varepsilon(P)}\right] + \frac{B}{1+\varepsilon} \ln\left[\frac{(S_0)}{(S_0) - (1+\varepsilon)(P)}\right], \tag{19}$$

where

$$K_1 = \frac{k_{-1} + k_2 + k_4}{k_1}, \qquad K_3 = \frac{k_{-3}}{k_3}, \qquad \varepsilon = \frac{k_4}{k_2},$$

$$A = \frac{K - \sigma\varepsilon}{(1 + \varepsilon)(E_0) - \varepsilon(S_0)},$$

$$B = \frac{(E_0)\,\sigma(1 + \varepsilon) - (S_0)\,K}{(E_0)(1 + \varepsilon)(E_0) - \varepsilon(S_0)},$$

$$K = 1 + \varepsilon - \frac{K_1}{K_3},$$

$$\sigma = K_1 + (S_0)\,.$$

The values of the parameters were estimated by a non-linear least squares minimization using numerical integration and optimization of the parameter[19]. The values of the kinetic parameters obtained were $k_2 = 9.05$, $k_4 = 0.235$, $K_1 = 40.05$, and $K_3 = 2.40$. Howell and Mangat[19] used Eq. (19) to predict the progress of hydrolysis under the various experimental conditions and reported that the single enzyme model based on competitive product inhibition by cellobiose and a first-order deactivation of the enzyme-substrate complex predicted the kinetic behavior of the hydrolysis reaction fairly well for an initial substrate concentration of 50 g l^{-1} and the time period up to around 90 h. The kinetic model by Howell and Mangat[19] appears to be applicable up to 3 or 4 days of duration, in the latter part of which the substrate multiplicity strongly affects the kinetics.

3.3 McLaren's Approach to Enzyme Reactions in a Structurally Restricted System

McLaren[39] studied the enzyme kinetics of the hydrolysis of insoluble substrates by various types of enzymes. He derived kinetic equations for such reaction systems based on the assumption that a soluble enzyme is adsorbed on the surface of a substrate in accordance with the Gyani-Freundlich isotherm and that the rate of digestion is proportional to the amount of the adsorbed enzyme. Thus,

$$E + A_s \rightleftarrows EA_s\,, \tag{20}$$

where
A_s = surface area of the substrate
E = enzyme concentration.
He also assumed that the solute is distributed between phase I (liquid) and phase II (solid) according to the Gyani-Freundlich equation, i.e.,

$$C_{II} = KC_I^n\,, \tag{21}$$

where
C_I = concentration in phase I
C_{II} = concentration in phase II
K = partition coefficient.

If phase II is a solid capable of only surface sorption, the surface concentration is given by

$$\frac{E_a}{A_s} = K\left(\frac{E_{\mathrm{I}}}{V_{\mathrm{I}}}\right)^{2/3} = KC^n , \tag{22}$$

where
E_a = number of moles of enzyme adsorbed on the surface, A_s, of phase II
E_{I} = number of moles of enzyme in the volume V_{I} of phase I at equilibrium
V_{I} = volume of phase I at equilibrium

McLaren[39] further assumed that

$$V_{\mathrm{initial}} = k'(EA_s) .$$

For a unit volume of a dilute suspension of the substrate with a unit surface area, this expression becomes

$$v_i = k'A_sK(E)^n/V_{\mathrm{I}} = k''K(E)^n . \tag{23}$$

Since

$$(E_0) = (E) + (EA_s) ,$$

Eq. (23) becomes

$$v_i = k''K[(E_0) - (EA_s)]^n . \tag{24}$$

This shows that unlike the Michaelis-Menten kinetics, the initial rate of hydrolysis for an insoluble substrate system is not a linear function of the total enzyme in the system because n is less than unity. It is, however, proportional to the surface area of the substrate in contact with the enzyme. The equation has been successfully applied to the action of hydrolytic enzymes on starch granules, cellulose, protein gels, and oil emulsions. McLaren[39] plotted the data obtained by Whitaker[59] for the cellulose hydrolysis as shown in Fig. 7. The equation is in good agreement with the data when $n = 0.66$ for cellulose and when $n = 0.77$ for the swollen linters. McLaren[39] postulated that the difference in the value of n stems from the difference of penetrations through the substrates; a larger value of n for the swollen linter indicates that the enzyme penetrates farther into the swollen linters.

Note that in McLaren's theory for the rate of enzyme action, K and n are not arbitrary constants but are determined from the adsorption isotherms. Thus, the rate constant, K, may also be calculated from appropriate rate data. Strictly speaking, the Gyani equation is applicable only to a solute of small molecules. Therefore, we may use a theory of the adsorption of macromolecules of the protein type on surfaces which is more sophisticated than the Gyani equation. In summary, the adsorptivity of enzymes on substrates is an important factor in kinetic analysis of cellulose hydrolysis. Unlike the Michaelis-Menten kinetics, the initial rate is not a linear function of the total enzyme concentration in the system. It is, however, proportional to the surface area of the substrate in contact with the solution.

McLaren[39] and McLaren and Packer[40] also derived an expression similar to Eq. (24), based on the assumption that the relation between E_0 and (E) is in the

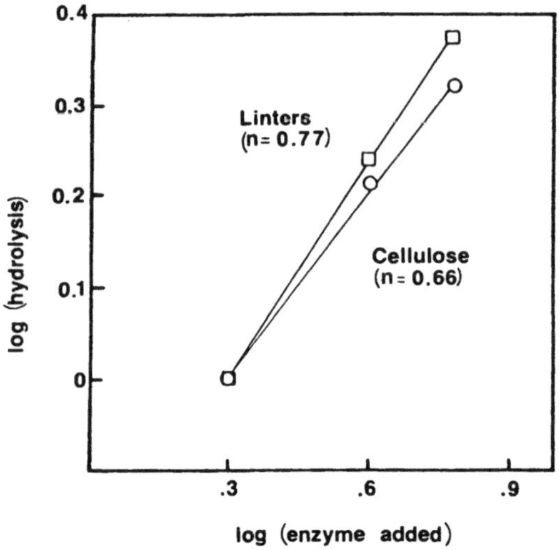

Fig. 7 Digestion of cellulose and swollen linters by cellulase[39]

form of the Langmuir adsorption isotherm. McLaren[39] was one of the early investigators who attempted to describe this structurally restricted cellulose-cellulase system, which takes into account the adsorption of enzyme onto the surface of the substrate and the hydrolysis by the adsorbed enzyme. However, the applicability of McLaren's approach is limited only to very simple situations because it ignores the substrate multiplicity, product inhibition and mode of action of cellulase. The more advanced models, which consider the adsorption and Michaelis-Menten kinetics, have been developed subsequently and are described in the following section.

3.4 Kinetic Models Based on the Combination of the Enzyme Adsorption Equation and the Michaelis-Menten Type Equation

Huang[21,22,23] carried out a series of kinetic studies on insoluble cellulose-cellulase systems. His kinetic models combine an enzyme adsorption equation and an equation of the Michaelis-Menten type. He[21] first investigated the enzymatic hydrolysis of an insoluble amorphous cellulose (Walseth cellulose) by *Trichoderma viride* cellulase, which contained high levels of C_1 and C_x, and a low level of β-glucosidase. This system can be considered basically as a one-substrate, one-enzyme catalyzed reaction. Huang assumed that the enzyme is rapidly adsorbed onto solids on contact and then gradually returns to the liquid phase as the reaction proceeds. Furthermore, end products, cellobiose and glucose were considered to be competitive inhibitors of cellulase action. Huang[21] depicted the mechanisms of the enzyme reaction as

$$E + S \underset{k_{-1}}{\overset{k_1}{\rightleftharpoons}} X_1 , \tag{25}$$

$$X_1 \xrightarrow{k_2} E + P , \tag{26}$$

$$E + P \underset{k_{-3}}{\overset{k_3}{\rightleftharpoons}} X_2 . \tag{27}$$

He further assumed that the cellobiose-to-glucose ratio remains reasonably constant, the Langmuir adsorption model prevails, the second reaction is the rate-determining step, the steady state is attained, and the concentration of adsorbed enzyme, X_1, is much smaller than its saturated maximum value (X_{1m}). Based on these assumptions, he derived the following rate equation:

$$v = \frac{d(P)}{dt} = \frac{k_2 X_{1m} K_1 (E)_0 (S)}{1 + K_3(P) + K_1(E)_0 + K_1(S)(X_{1m} - X_1)}, \tag{28}$$

where

$$K_1 = \frac{k_1}{k_{-1}},$$

$$K_3 = \frac{k_3}{k_{-3}}.$$

Huang[21] substituted $(S) = (S)_0 - (S)$ into Eq. (28) and obtained the following expression upon integration of the resultant equation:

$$\frac{1}{\bar{v}} = \frac{1 + K_1(E)_0 + K_3(S)_0}{K_2 X_{1m} K_1(E)_0} \frac{1}{(\bar{S})} + \frac{X_{1m} K_1 - K_3}{k_2 X_{1m} K_1(E)_0}, \tag{29}$$

where

$$\frac{1}{(\bar{S})} = \frac{1}{(P)} \ln \frac{(S)_0}{(S)_0 - (P)},$$

$$\bar{v} = \frac{(P)}{t}.$$

Huang[21] plotted the results of the hydrolysis experiments with amorphous cellulose according to Eq. (29), $1/\bar{v}$ against $1/(\bar{S})$. The experiments were carried out at several initial substrate concentrations and three initial enzyme concentrations. Table 1 shows the experimental and calculated values of the slopes and intercepts of the resultant plots for different values of $(S)_0$ and $(E)_0$. Note that at two enzyme levels, the data of $1/\bar{v}$ versus $1/(\bar{S})$ show sufficient linearity. Thus, the data permit the

Table 1. Experimental and calculated slopes and intercepts according to Eq. (29)[21]

$(E)_0$ mg ml^{-1}	$(S)_0$ mg ml^{-1}	Experimental		Calculated	
		slope	intercept	slope	intercept
0.38	10.0	3.68	−0.33	3.69	−0.33
	7.2	2.68	−0.30	2.69	−0.33
	5.0	1.90	−0.32	1.90	−0.33
	3.6	1.38	−0.36	1.40	−0.33
0.25	10.7	6.00	−0.50	5.97	−0.50
	7.1	4.02	−0.51	4.01	−0.50
	5.2	2.98	−0.51	2.97	−0.50
	3.5	2.05	−0.54	2.05	−0.50
0.15	5.0	4.91	−0.88	4.75	−0.83
	3.8	3.68	−0.81	3.66	−0.83

evaluation of kinetic parameters k_2, K_1, X_{1m}, and subsequently K_3; they are 80.3 h^{-1}, 1.68 ml mg^{-1}, 0.305 mg mg^{-1}, and 5.61 ml mg^{-1}, respectively. Calculated data based on these parameters agree satisfactorily with the experimental data up to 70% conversion.

Enzymatic degradation of usual cellulose such as Solka Floc would require a model consisting of two substrates, amorphous and crystalline cellulose. To apply the kinetic model for Solka Floc, Huang[22] transformed the kinetic expression of Eq. (29) into the following form.

$$\bar{v} = \frac{(P)}{t} = \frac{k_2 X_{1m} K_1 (E)_0}{X_{1m} K_1 - K_3} - \frac{1 + K_1(E)_0 + K_3(S)_0}{X_{1m} K_1 - K_3} \cdot \left\{ \frac{1}{t} \ln \frac{(S)_0}{(S)} \right\}. \tag{30}$$

He plotted the batch reaction data from the hydrolysis of Solka Floc, which contained both the amorphous and the crystalline portions; specifically, he plotted

$\frac{(P)}{t}$ against

$$\left\{ \frac{1}{t} \ln \frac{(S)_0}{(S)} \right\}.$$

He observed linearity for all three enzyme levels at several solid suspensions and interpreted that the linearity results from hydrolysis of the amorphous fraction of cellulose, whose concentration remains fairly constant throughout a batch run (up to 30 h of reaction time).

It is generally agreed that the crystalline cellulose is more resistant to degradation than the amorphous portion because hydrogen bonds are present[6,10]. Thus, a two-substrate system must be considered in kinetic modeling of the hydrolysis of cellulose, e.g., Solka Floc, which contains both amorphous and crystalline states. In addition, adsorption of cellulase onto solids and the effect of product inhibition should be included in analyzing this heterogeneous enzyme-catalyzed reaction. Huang[23] also constructed a kinetic model stipulating that enzyme E be adsorbed on crystalline cellulose S_c to form complex X_1, which, in turn, yields amorphous cellulose S_a. The enzyme is adsorbed on amorphous cellulose S_a to form complex X_2, which is degraded to soluble product P. The product then combines with the enzyme to form complex X_3, which inhibits the enzyme action. This mechanism can be depicted as

$$E + S_c \underset{k_{-1}}{\overset{k_1}{\rightleftharpoons}} X_1 \underset{k_{-2}}{\overset{k_2}{\rightleftharpoons}} E + S_a, \tag{31}$$

$$E + S_a \underset{k_{-3}}{\overset{k_3}{\rightleftharpoons}} X_2 \xrightarrow{k_4} E + P, \tag{32}$$

$$E + P \underset{k_{-5}}{\overset{k_5}{\rightleftharpoons}} X_3. \tag{33}$$

Based on an assumption similar to that used in his earlier papers[21,22], Huang[23] derived the following expression for the rate of reaction, v.

$$v = \frac{d(P)}{dt} = \frac{k_4 X_m K(E)_0 (S) f}{1 + K_5(P) + K(E)_0 + K X_m(S)}, \tag{34}$$

where

f = fraction of amorphous cellulose in the total cellulose,
X_m = saturation amount of enzyme adsorbed per unit weight of cellulose,

$$K = \frac{k_1}{k_{-1}} = \frac{k_3}{k_{-3}},$$

$$K_5 = \frac{k_5}{k_{-5}}.$$

By assuming that f is constant and that $(S) = (S)_0 - (P)$, integration of Eq. (34) gives

$$t = \frac{1 + K(E)_0 + K_5(S)_0}{k_4 X_m K(E)_0 f} \ln \frac{(S)_0}{(S)} + \frac{X_m K - K_5}{k_4 X_m K(E)_0 f}(P), \tag{35}$$

which can be rearranged to a linear form:

$$\frac{(P)}{t} = \frac{K_4 X_m K(E)_0 f}{X_m K - K_5} - \frac{1 + K(E)_0 + K_5(S)_0}{X_m K - K_5}\left\{-\frac{1}{t}\ln\frac{(S)}{(S)_0}\right\}. \tag{36}$$

Thus, plotting

$$\frac{(P)}{t}$$

against

$$\left\{-\frac{1}{t}\ln\frac{(S)}{(S)_0}\right\},$$

by keeping $(E)_0$ constant and varying $(S)_0$, should yield a family of straight lines. The slope is a function of $(S)_0$ expressed in Eq. (36) as confirmed by Huang's experimental data[23]. Table 2 lists the slopes and intercepts obtained from his data.

Table 2. Slopes and intercepts of Solka Floc BW-200 hydrolysis from a plot according to Eq. (36)[22,23]

$(E)_0$ mg ml^{-1}	$(S)_0$ mg ml^{-1}	Slope	Intercept
0.76	176	175	−0.115
	111	110	−0.137
	41.7	40.5	−0.171
	20.4	19.7	−0.167
	14.2	14.1	−0.180
0.44	111	111	−0.068
	41.7	41.7	−0.131
	20.4	20.8	−0.185
0.25	111	111	−0.019
	41.7	41.5	−0.046
	20.4	20.6	−0.085

However, the intercepts at a given $(E)_0$ were not identical, which, according to Eq. (36), would result from different values of f for different values of $(S)_0$. In fact, it was observed that f decreases as $(S)_0$ increases. Huang[23] concluded that this is reasonable because less enzyme becomes available for swelling per unit weight of crystalline cellulose as $(S)_0$ increases, and this would result in a small value of f. Therefore, he postulated that the negative intercepts results from the strong product inhibition.

Kim[26] also attempted to model and simulate cellulose hydrolysis with *T. viride* cellulase. The model takes into account such plausible mechanisms as bi-composition of cellulose, negligible mass transfer resistances, cellulase adsorption, formation and decomposition of the cellulose-cellulase complex, product inhibition, and cellulase deactivation. The assumptions were that cellulose fiber is of long cylindrical form, the concentration of cellulase adsorbed on cellulose is proportional to its concentration in the aqueous solution, and the adsorbed cellulase decays exponentially. The mechanism as suggested can be shown as follows:
Complex formation and decomposition:

$$C_A + E_A^A \underset{k_{-1}}{\overset{k_1}{\rightleftharpoons}} X_A \overset{k_2}{\longrightarrow} G_X + E_A^A, \tag{37}$$

$$C_C + E_A^C \underset{k_{-3}}{\overset{k_3}{\rightleftharpoons}} X_C \overset{k_4}{\longrightarrow} C_A + E_A^C. \tag{38}$$

Inhibition:

$$X_A + G_X \underset{k_{-5}}{\overset{k_5}{\rightleftharpoons}} Y_A, \tag{39}$$

$$X_C + G_X \underset{k_{-6}}{\overset{k_6}{\rightleftharpoons}} Y_C. \tag{40}$$

Enzyme deactivation:

$$E_A^A \overset{k_7}{\longrightarrow} E_A^d, \tag{41}$$

$$E_A^C \overset{k_7}{\longrightarrow} E_A^d, \tag{42}$$

where

C_A, C_c = concentrations of amorphous and crystalline cellulose, respectively
E_A^A = concentration of the enzyme active on amorphous cellulose
E_A^C = concentration of the enzyme active on crystalline cellulose
X_A, X_c = enzyme-cellulose complexes
G_x = products
Y_A, Y_c = enzyme-substrate-product complexes
E_A^d = deactivated form of the enzyme.

Table 3. Rate equations for the hydrolysis of bi-composition cellulose[26]

Rate Equations	
$\dot{C}_A = -PA + \alpha X_c$	
$\dot{C} = -PC$	
$\dot{G}_x = \dot{\gamma} X_A - QA - QC$	
$\dot{X}_A = \gamma X_A + PA - QA$	$\dot{X}_c = \alpha X_c + PC - QC$
$\dot{Y}_A = QA$	
$\dot{Y}_c = QC$	
where	
$PA = \frac{E_{A0}}{C_0^{1/2}} \frac{C^2 A}{C^{1/2}} \exp(-\varepsilon\tau) - X_A/P_{k1}$	
$PC = \beta\left[\frac{E_{A0}}{C_0^{1/2}} \frac{C_c^2}{C^{1/2}} \exp(-\varepsilon\tau) - X_c/P_{k3}\right]$	
$QA = \delta(X_A G_x^m - Y_A/P_{k5})$	
$QC = \eta(X_C G_x^m - Y_c/P_{k6})$	
$\bullet \equiv \frac{d}{d_\tau} \quad \tau = k_1 t \quad t = \text{time}$	
$\alpha = \frac{k_4}{k_1} \quad \beta = \frac{k_3}{k_1} \quad \gamma = \frac{k_2}{k_1} \quad \delta = \frac{k_5}{k_1} \quad \eta = \frac{k_6}{k_1} \quad \varepsilon = \frac{k_7}{k_1}$	

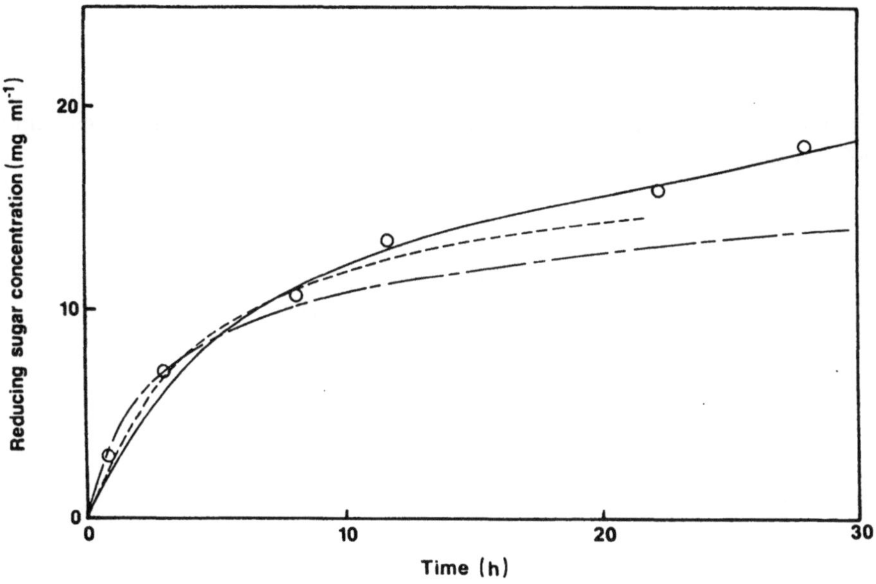

Fig. 8 Comparison between exact and approximate solutions, the initial rate (INI) and the steady-state kinetics (SS)[26]

The common type of product inhibition in an enzyme reaction results from the combination of products and free enzymes. Kim[26] postulated that this type of inhibition is not likely to affect cellulose hydrolysis since this system is heterogeneous, and thus, the enzyme adsorbed on the cellulose matrix is primarily responsible for the reaction. He also postulated that the most probable inhibitory mechanism results from a combination of the enzyme-substrate complexes and the products. Based on these postulations, a set of rate equations was derived as shown in Table 3. These equations were solved by using simulation computer subroutines such as CSMP, MIMIC and GELG. Fig. 8 compares the exact and approximate solutions, the initial rate (INI), and steady-state kinetics (SS). Kim[26] pointed out that a simple Michaelis-Menten type kinetics does not adequately represent the system, and the steady-state solution usually obtained is invalid because of the complicated nature of the cellulose-cellulase system.

3.5 Kinetic Model Based on the Depolymerization Mechanism

Several kinetic models have been proposed, which take into account the mode of action of different cellulase components on molecules with different degrees of polymerization. Suga et al.[53] considered the degradation of polysaccharide chains by either an endo-enzyme or an exo-enzyme alone and also by various combinations of endo-enzymes and exo-enzymes. They assumed that the reaction between the enzyme and substrate is to be the common Michaelis-Menten type reaction, i.e.,

$$S + E \underset{k_{-1}}{\overset{k_1}{\rightleftarrows}} ES \xrightarrow{k_3} P + E \,. \tag{43}$$

By resorting to a quasi-steady state assumption, they obtained the following equation for the rate of change in the concentration of a cellulose fragment with chain length $i(C_i)$ in the presence of an endo enzyme alone:

$$\frac{\mathrm{d}C_i}{\mathrm{d}t} = -\frac{k_3 C_{Et}(i-1)\,C_i}{K_m + \sum\limits_{j=2}^{\infty}(j-1)\,C_j} + 2\,\frac{k_3 C_{Et} \sum\limits_{j=i+1}^{\infty} C_j}{K_m + \sum\limits_{j=2}^{\infty}(j-1)\,C_j} \,. \tag{44}$$

where

C_{Et} = total enzyme concentration

When only an exo-enzyme, which produces selely monomer (C_1), is present, the rate of change of C_1 is given by

$$\frac{\mathrm{d}C_1}{\mathrm{d}t} = -\frac{k_3' C_{Et}' C_i}{K_m' + \sum\limits_{j=2}^{\infty} C_j} + \frac{k_3' C_{Et}' C_{i+1}}{K_m' + \sum\limits_{j=2}^{\infty} C_j} \,. \tag{45}$$

For the degradation of cellulose with various combinations of an endo-enzyme and exo-enzyme, the following equation was obtained:

$$\frac{dC_i}{dt} = - \frac{k_3 C_{Et}(i-1)\, C_i}{K_m + \sum_{j=2}^{\infty} (j-1)\, C_j} + 2 \frac{k_3 C_{Et} \sum_{j=i+1}^{\infty} C_j}{K_m + \sum_{j=2}^{\infty} (j-1)\, C_j} - \frac{k_3' C_{Et}' C_i}{K_m' + \sum_{j=2}^{\infty} C_j} + \frac{k_3' C_{Et}' C_{i+1}}{K_m' + \sum_{j=2}^{\infty} C_j} . \quad (46)$$

Suga et al.[53] solved Eqs. (44), (45) and (46) to obtain the molecular weight distributions of cellulose fragments during the progress of hydrolysis. This model does not consider branch points of linkages, inhibition by products, and diffusion and adsorption of the enzyme through and on the substrate. However, the synergistic effect between the endo-enzyme and exo-enzyme can be evaluated on the basis of this model.

While the model proposed by Suga et al. considers only the endo-enzyme and exo-enzyme, the enzyme complex involved in cellulose hydrolysis contains endo-glucanase, exo-glucanase, cellobiosylhydrolase, and cellobiase[32]. The model of Suga et al. has been further extended by Lee et al.[30,31] to the cellulase system which has four enzyme components. The extended model includes the action of cellobiosylhydrolase and cellobiase; therefore, it describes more completely the action of the cellulase complex than the previous model[53].

Okazaki and Moo-Young[43] proposed another extended model based on the Michaelis-Menten type kinetics for concurrent random and endwise attack of the substrate involving end-product inhibitions and three types of enzymes: endo-glucanase, cellobiosylhydrolase and β-glucosidase. They used the model to investigate the synergistic effect among the cellulase components, the dependency of the enzyme activity on the degree of polymerization of cellulose, and the substrate inhibition in cellulose degradation. Those interested in the mathematical details of the model may refer to the original paper[43].

These depolymerization models are suitable for expressing the mode of action of each of the cellulase components and the synergistic action among them. In spite of this merit, each type of the depolymerization models contains a relatively large number of parameters. Also these models ignore some significant factors in kinetic interactions, such as the effect of the structural features of cellulose, mass transfer of enzyme and products, and the adsorption of enzyme.

3.6 Distributed Parameter Kinetic Models

Ross and Updegraff[48] proposed a combination of Fickian diffusion and Michaelis-Menten (M-M) kinetics to describe the rate of a diffusion-coupled biochemical reaction, specifically, the conversion of cellulose to protein by *Myrothecium verrucaria*. In their model, Fick's second law is assumed to govern the diffusion, and the differential equation of the M-M rate mechanism is assumed to govern the reaction. Thus, the equation

$$\frac{\partial C}{\partial t} = D\frac{\partial^2 C}{\partial x^2}, \tag{47}$$

is obtained, which is governed by the following boundary conditions (B.C.) and initial conditions (I.C.);

$$\text{B.C.;} \quad -D\frac{\partial C}{\partial x} = k\frac{C}{K + C} \quad \text{at } x = 0, \tag{48}$$

$$C = 0 \quad \text{at } x \to \infty, \tag{49}$$

$$\text{I.C.;} \quad C = 0 \quad \text{at } x = 0, \tag{50}$$

where C is the concentration of the product. This non-linear problem was solved by an approximate method (perturbation method) developed by Ross. When the principal interest is the product (protein) formation, the time integral of the product flux at the cellulose surface is obtained as

$$M = -\int_0^t D\left(\frac{\partial C}{\partial x}\right)\bigg|_{x=0} dt = kt - K\sqrt{\pi D t}. \tag{51}$$

This equation provides the basis for correlating experimental results. Ross and Updegraff[48] plotted experimental data obtained by Updegraff[55] in two ways, namely, Mt^{-1} *vs* $\sqrt{t}^{-1}$ and $M\sqrt{t}^{-1}$ *vs* $\sqrt{t}$. These plots yielded two linear correlations, one with a slope of $-K\sqrt{\pi D}$ and an intercept of k and the other with a slope of k and an intercept of $-K\sqrt{\pi D}$. Such correlations permitted the identification of regions of diffusion control and reaction control. Since the model, as represented by Eqs. (47) and (51), is an oversimplification for all but the most elementary microbial processes, it can represent only very simplified kinetics.

Suga et al.[52] developed a model for an enzymatic reaction on an insoluble substrate. They investigated the kinetic behaviour of the enzymatic breakdown of cross-linked dextran by a dextranase from *Penicillium funiculosum* as a model for the enzymatic degradation of insoluble materials. This system is simpler than the cellulose-cellulase system because only one enzyme is involved[58]. For a reaction between an insoluble substrate and an enzyme, the mass transfer of the enzyme through the cross-linked substrate should be considered. Suga et al.[52] assumed that the reaction is of the common Michaelis-Menten type.

$$S + E \underset{k_2}{\overset{k_1}{\rightleftharpoons}} ES \overset{k_3}{\longrightarrow} P + E. \tag{52}$$

They also made the following assumptions: (1) diffusion of the enzyme and the released substrate fragments can be represented by Fick's first law, (2) the diffusion coefficient of the soluble substrate fragments can be represented by a function of their average molecular weight, (3) the average pore size of the particles is constant, (4) the restricted diffusion coefficient of the enzyme and that of the released substrate fragments follow the equation proposed by Renkin[47], and (5) the Sephadex particles are spherical.

According to the model by Suga et al.[52], a material balance for the substrate fragments of chain length i over the differential volume of a spherical particle gives

$$\frac{\partial C_i}{\partial t} = \frac{1}{r^2}\frac{\partial}{\partial r}\left(D_s r^2 \frac{\partial C_i}{\partial r}\right) + R(C_i, C_E) = D_s \frac{\partial^2 C_i}{\partial r^2} + D_s \frac{2}{r}\frac{\partial C_i}{\partial r} + \frac{\partial D_s}{\partial r}\frac{\partial C_i}{\partial r} + R(C_i, C_E), \tag{53}$$

where

C_i = substrate fragments of chain length i,
D_s = average restricted diffusion coefficient of soluble substrate fragments,
C_E = concentration of the enzyme in particles.

By assuming random fission and constant values for k_3 and K_m for all chain lengths,

$$R(C_i, C_E) = -\frac{k_3 C_E (i-1) C_i}{K_m + \sum_{j=2}^{\infty}(j-1) C_j} + 2\frac{2k_3 C_E \sum_{j=i+1}^{\infty} C_j}{K_m + \sum_{j=2}^{\infty}(j-1) C_j}. \tag{54}$$

Combining Eqs. (53) and (54) yields

$$\frac{\partial C_s}{\partial t} = D_s \frac{\partial^2 C_s}{\partial r^2} + D_s \frac{2}{r}\frac{\partial C_s}{\partial r} + \frac{\partial D_s}{\partial r}\frac{\partial C_s}{\partial r} - \frac{k_3 C_E C_s}{K_m + C_s}, \tag{55}$$

where C_s is the total sum of the concentrations of breakable substrates. The material balance for the enzyme over the differential volume of a spherical particle gives

$$\frac{\partial C_E}{\partial t} = D_E\left(\frac{\partial^2 C_E}{\partial r^2} + \frac{2}{r}\frac{\partial C_E}{\partial r}\right). \tag{56}$$

The restricted diffusion coefficient as proposed by Renkin[47] is

$$\phi = \frac{D_E}{D_{E0}} = \left(1 - \frac{a}{\gamma}\right)^2 \left\{1 - 2.104\left(\frac{a}{\gamma}\right) + 2.09\left(\frac{a}{\gamma}\right)^3 - 0.95\left(\frac{a}{\gamma}\right)^5\right\}, \tag{57}$$

where

a = radius of enzyme,
γ = pore radius.

Also, the restricted diffusion coefficient of released substrate fragments, D_s, is expressed as

$$D_s = D_{SF}\eta, \tag{58}$$

where

D_{SF} = average free diffusion coefficient of the fragments according to their average molecular weight, $cm^2\ s^{-1}$.

Furthermore, the correction factor pertaining to diffusional restriction, η, is defined as

$$\eta = b\phi \,, \tag{59}$$

where b is an empirical constant. A material balance for the substrate in the solution gives

$$V_L \frac{dC_{sL}}{dt} = 4\pi R^2 n\bar{k}_S(C_{s,r=R} - C_{sL}) - V_L R(C_{sL}, C_{E_0}) \,, \tag{60}$$

where

V_L = volume of the liquid phase,
C_{sL} = concentration of the substrate in solution,
R = particle radius,
n = number of particles,
$\bar{k}_S$ = overall mass transfer coefficient of the substrate.

Suga et al.[52] solved these equations numerically, subject to the appropriate boundary and initial conditions, and determined the parameters of the model, such as K_m, V, K_d, D_{E0}, D_{SF}, D_{ES}, a and D_s experimentally. They verified their model against experimental data from the degradation of the cross-linked dextran. Fig. 9 compares the experimental data with the theoretical curves for the following values of parameters:

$K_m = 0.952 \times 10^{-6}$ mol cm^{-3}, $V = 0.222 \times 10^{-8}$ mol cm^{-3},
R(particle radius) $= 0.01$ cm, $b = 0.009$, $a = 0.283 \times 10^{-6}$ cm,
$\gamma = 0.560 \times 10^{-6}$ cm, $c = 0.001$ cm.

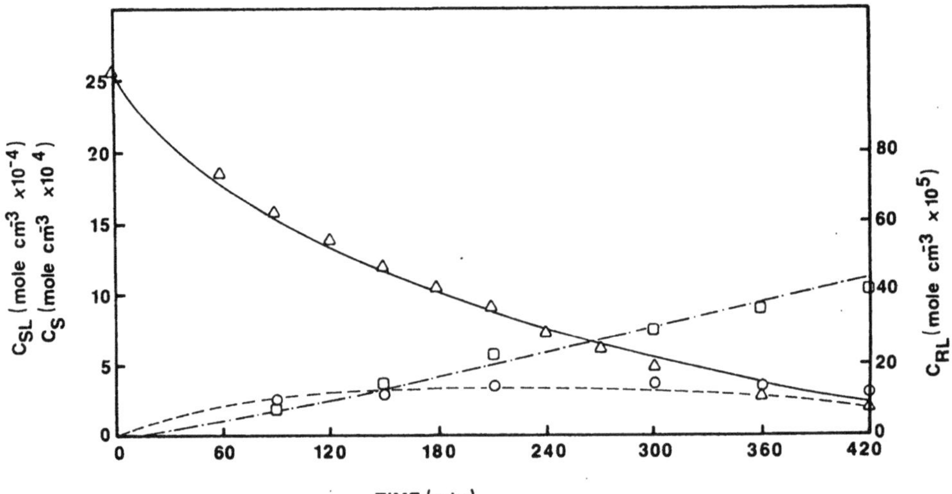

Fig. 9 Comparison between experimentally obtained change in substrate concentration in particle (△) and in solution (□), and concentration of reducing sugar (○) and theoretical curves[52]

The experimental data agree well with the theoretical surves. Thus, the model can be used to explain the behavior of the enzymatic breakdown of cross-linked dextran when the kinetics of the reaction and the physical properties of the substrate are known. This type of model may also be useful in the interpretation of the behaviour for the enzymatic degradation of other insoluble substrates such as cellulose.

4 Kinetics of the Reactions of Cellulases with β-1,4-Oligoglucosides

Whitaker[59,60] studied the degradation of oligoglucosides containing two, three, four, five, and six glucose units by *Myrothecium verrucaria* cellulase, in order to compare the rates of hydrolysis and sequences of products formed from them. Table 4 shows the turnover numbers and rate constants at different initial concentrations of each

Table 4. Initial turnover numbers and rate constants at constant enzyme concentration[60]

Substrate	Initial conc. mM	Turnover number	Rate constants: Zero-order mole ml^{-1} min^{-1} $\times 10^9$	Rate constants: First-order $min^{-1} \times 100$
Cellobiose	28.2	5.1	1.6	
	14.4	5.3	1.7	
	7.4	6.0	1.9	
	3.7	6.2	2.1	
Cellotriose	15.5	210	67	0.50
	10.2	129	41	0.48
Cellotetraose	4.0	49	16	0.45
	14.3	377	121	0.98
	7.4	316	102	1.78
	3.6	384	123	4.75
Cellopentaose	8.4	460	147	
	5.3	435	139	
	2.7	473	152	
Cellohexaose	4.5	430	138	
	2.2	439	141	

different cellooligosaccharide obtained. Turnover numbers reveal two marked trends. First, the turnover numbers increase with an increasing degree of polymerization (DP) of the substrate up to at least a DP of five. The rate of increase is particularly high when the DP is between two and four. Second, the turnover numbers are inversely proportional to the initial substrate concentration. The initial velocity constants reflect the same trend as the turnover numbers.

A theoretical treatment of the kinetics of the degradation of long-chain polymers was given by Simha[49]. To test the applicability of this type of analysis, Whitaker[60,62] compared his experimental results with the calculated values obtained from the following rate equations:

$$\frac{dG^4}{dt} = -k^4 G^4 = -(k^4_{2:2} + k^4_{3:1})\, G^4 , \tag{61}$$

$$\frac{dG^3}{dt} = \frac{3}{4} k^4_{3:1} G^4 - k^3 G^3 , \tag{62}$$

$$\frac{dG^2}{dt} = k^4_{2:2} G^4 + \frac{2}{3} k^3 G^3 - k^2 G^2 , \tag{63}$$

where

G^2, G^3, G^4 = concentration of cellobiose, cellotriose, cellotetraose, respectively
k^3, k^4 = first-order reaction rate constants for cellotriose and cellotetraose

To simplify the form of the solution, the rate constant for cellobiose, k^2, is taken to be that of a zero-order reaction. Under the conditions $t = 0$, $G^4 = G^4_0$, $G^3 = G^2 = G^1 = 0$ and $k^4 \neq k^3$, the solution of Eqs. (61)–(63) are as follows:

$$G^4 = G^4_0\, e^{-k^4 t} , \tag{64}$$

$$G^3 = \frac{3}{4}\left[\frac{k^4_{3:1}}{k^4 - k^3}\right] G^4_0 (e^{-k^3 t} - e^{-k^4 t}) , \tag{65}$$

$$G^2 = \frac{k^4_{2:2}}{k^4} G^4_0 (1 - e^{-k^4 t}) + \frac{k^4_{3:1}}{2(k^4 - k^3)} G^4_0 (1 - e^{-k^3 t}) - \frac{k^4_{3:1} k^3}{2k^4(k^4 - k^3)} G^4_0 (1 - e^{-k^4 t}) - k^2 t , \tag{66}$$

$$\Sigma G_i = G^4_0 , \tag{67}$$

$$G^1 = G^4_0 - G^4 - G^3 - G^2 . \tag{68}$$

The calculated values agree reasonably well with the experimental values which suggests that the conditions expressed by the rate equations are approximately satisfied during the degradation considered.

Later Whitaker[61,62] attempted to define the kinetic behaviour of the action of cellulase on β-1,4-oligoglucosides. He found that, at sufficiently low cellulase concentration, (E) and high substrate concentration, (S), the rate of hydrolysis is directly proportional to the concentrations of both the cellulase and substrate, i.e.,

$$-\frac{dS}{dt} = k(E)\,(S) , \tag{69}$$

where k is the second-order rate constant. When (E) is constant, this expression reduces to

$$-\frac{dS}{dt} = k'(S) , \tag{70}$$

where k' is the first-order or, strictly speaking, the pseudo first-order rate constant. The rate constants obtained are listed in Table 5. It has also been observed that the enzymatic hydrolysis of β-1,4-oligoglucosides is inhibited by increasing the concen-

Table 5. Pseudo first-order rate constants of the hydrolysis of β-1,4-oligoglucosides by *Myrothecium* cellulase (20 μg of protein in ml)[62]

Substrate	K', $s^{-1} \times 10^5$
Cellobiose	1.2
Cellotriose	16
Cellotetraose	83
Cellopentaose	500

tration of the substrate, irrespective of the chain length of the latter, as shown in Fig. 10. The inhibition by the substrate might have resulted from an ESS complex formation. (A similar observation is reported in another work[42] where cellulase components from the culture filtrate of *Irpex lacteus* have been employed).

Li et al.[33] estimated the Michaelis constante (K_m) of the endo-glucanase and exo-glucanase from *T. viride* for the cellodextrin series, as shown in Table 6. The Michaelis constants (K_m) of exo-glucanase are much smaller than those of the endo-glucanase for all polymer series, cellobiose through cellohexaose. This means that the exo-glucanase, which successively removes single glucose units from the non-reducing end of a cellulose chain, is more like β-glucosidase than endo-glucanase.

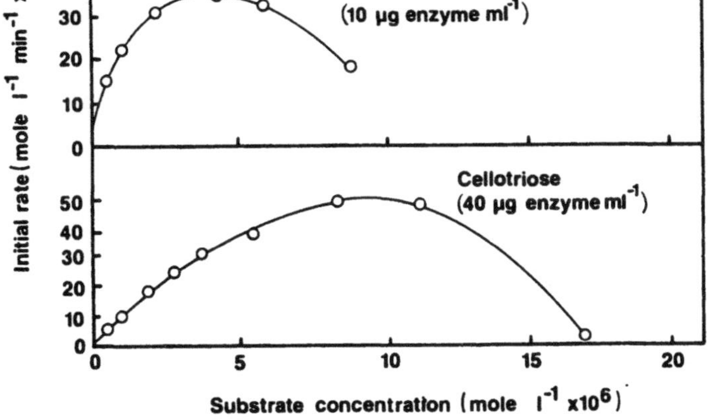

Fig. 10 Rates of hydrolysis of β-1,4-oligoglucosides by *Myrothecium* cellulase with increasing concentrations of substrates[62]

Table 6. Michaelis constants for the cellulose polymer series of endo- and exo-glucanase[33]

Substrate	Michaelis constant	
	Endo-glucanase	Exo-glucanase[b]
Cellobiose	190×10^{-4} M[a]	220×10^{-5} M[a]
Cellotriose	31×10^{-4} M	18×10^{-5} M[a]
Cellotetraose	28×10^{-4} M	6.5×10^{-5} M[a]
Cellopentaose	7.0×10^{-4} M	6.0×10^{-5} M
Cellohexaose	1.0×10^{-4} M[a]	16.0×10^{-5} M[a]

[a] These Lineweaver-Burk plots showed apparent substrate inhibition at concentrations exceeding 0.05 M

[b] It is not clear whether this enzyme was exo-β-1,4-glucanase (EC not given) or β-glucosidase (EC 3.2.1.21)

5 Kinetics of β-Glucosidase

Gong et al.[16] studied the purification, physical characteristics and mechanism of cellobiase from *T. viride*. Based on the analysis of the initial rate data, plotted in the form of the Lineweaver-Burk, they concluded that cellobiase is non-competitively inhibited by the product glucose. Therefore, the reaction sequence for this enzyme could be written as

$$E + GG \underset{k_2}{\overset{k_1}{\rightleftarrows}} E^* \xrightarrow{k_3} E + 2G\,, \tag{71}$$

$$E^* + G \underset{k_6}{\overset{k_5}{\rightleftarrows}} E^*G\,, \tag{72}$$

$$E + G \underset{k_8}{\overset{k_7}{\rightleftarrows}} EG\,, \tag{73}$$

where

E = free enzyme
GG = cellobiose
G = glucose.

By applying the pseudo-steady state assumption, Gong et al.[16] derived the inverted form of the kinetic expression.

$$\frac{1}{v} = \frac{K[1 + (G)/K_{i,2}]}{V} \times \frac{1}{(GG)} + \frac{1 + (G)/K_{i,1}}{V}\,, \tag{74}$$

where

$$K = \frac{k_2 + k_3}{k_1},$$

$$K_{i,1} = \frac{k_6}{k_5},$$

$$K_{i,2} = \frac{k_8}{k_7},$$

$$V = 2k_3E_{tot}.$$

Since the plots and replots are linear, values of the kinetic constants, K, $K_{i,1}$, and $K_{i,2}$ can be determined directly from them. The kinetic constants for three cellobiase fractions are summarized in Table 7. The kinetic equation derived from the proposed mechanism accurately predicted the time course of the hydrolysis of cellobiose by cellobiase over an eightfold range of the substrate concentration and at conversions of over 90% [16]. The results indicate that cellobiase from *T. viride* consists of several chromatographically distinct and yet kinetically similar fractions, which are subject to product inhibition, and that these fractions of cellobiase inhibit the hydrolysis of cellobiose according to a non-competitive mechanism.

Table 7. Values of kinetic constants for peaks 1, 2, and 3 cellobiase fractions[16]

Kinetic constant	Cellobiase enzyme fraction		
	Peak 1	Peak 2	Peak 3
$\frac{V}{E_{tot}}$ $\frac{\mu mol\ glucose}{min\ mg^{-1}\ protein}$	66.2	116.0	44.6
K (mM)	2.65	2.5	2.74
$K_{i,1}$ (mM)	—	16.4	14.7
$K_{i,2}$ (mM)	—	1.22	4.26

6 Summary

Enzyme kinetics have been concerned mainly with soluble substrates. Relatively few kinetic studies have involved soluble enzymes and insoluble substrates, e.g. enzymatic hydrolysis of cellulose. However, research on the enzymatic hydrolysis of cellulose to glucose has been gaining momentum; consequently, the mechanism and kinetics of cellulose hydrolysis by cellulase have begun to be understood; yet much remains to be investigated.

Because of the heterogeneous nature of the cellulose-cellulase reaction system, it shares only a few characteristics with the soluble substrate-enzyme reaction system. Some of the key mechanisms that distinguish the two systems include diffusion of the enzyme into the cellulose particle, adsorption and desorption of the enzyme on the cellulose surface, surface reactions and fragmentations, diffusion of the products, and product inhibition.

Since cellulose is an insoluble substrate, the rate of formation of an enzyme-substrate complex may be appreciably influenced by mass transfer resistance. However, the mass transfer resistance can be reduced substantially with adequate mixing. The cellulolytic enzyme is strongly adsorbed by cellulose particles under conditions optimal for the enzyme action, and the quantity of the enzyme adsorbed is proportional to the cellulase concentration. It appears that the Langmuir isotherm can be used to relate the concentration of the adsorbed enzyme to the initial free enzyme concentration. Some investigators have proposed that the amount of enzyme initially adsorbed onto cellulose is primarily responsible for the hydrolysis throughout the entire reaction, but this proposition needs further verification. The reaction between cellulose and cellulase is heterogeneous in character, and it has been suggested, therefore, that the reaction rate should be proportional to the extent of the cellulose surface that is accessible to the enzyme molecule. However, fragmentation and changes of the surface area during hydrolysis show that the enzyme action is not exclusively a surface phenomenon. Two product inhibition mechanisms have been reported: competitive inhibition and non-competitive product inhibition; however, neither has been verified definitely.

Several kinetic models have been proposed for the hydrolysis of insoluble cellulose catalyzed by cellulase. Some of them are purely empirical, and others are analogous to the Michaelis-Menten equation. Some of the models are based on the assumption that the enzyme is adsorbed on the surface of the substrate and that the rate of digestion is proportional to the concentration of the adsorbed enzyme. Kinetic models, based on a combination of enzyme adsorption and the Michaelis-Menten kinetics, have also been proposed. A distributed parameter model, which takes into account mass transfer and reaction, was proposed. However, the experimental work to verify the model uses cross-linked dextran rather than cellulose as the substrate. It appears that no satisfactory models have been proposed which are applicable to the hydrolysis of various forms of insoluble cellulose. To derive a mechanistic kinetic model, the structural features of cellulose and the mode of action of the cellulase complex should be fully investigated.

The results of the kinetic studies of the hydrolysis of soluble β-1,4-oligoglucosides catalyzed by cellulase may provide an approximate picture of the kinetics of the enzymatic hydrolysis of insoluble cellulose. Kinetic studies show that β-glucosidase is non-competitively inhibited by the product glucose.

Because of the heterogeneous nature of a cellulose-cellulase system, a distributed parameter model may be desirable. Such a model should take into account mass transfer, reaction, product inhibition, structural features of cellulose, and the mode of action of cellulase.

7 Acknowledgement

This is Department of Chemical Engineering contribution No. *79-326-J*. Agricultural Experiment Station, Kansas State University. This work also supported by the Department of Energy grant No. DE-FG02-79ET-00080 37.

8 Symbols

a	radius of enzyme, cm
A_s	surface area of the substrate, L^2
b	constant, dimensionless
C	concentration of the product [Eqs. (47) or (51)], ML^{-3}
C_A	concentration of amorphous cellulose, ML^{-3}
C_C	concentration of crystalline cellulose, ML^{-3}
C_E	concentration of enzyme in the particle, mol cm^{-3}
C_{Et}	total enzyme concentration, ML^{-3}
C_i	cellulose concentration of component i [Eq. (6)], ML^{-3}
C_i	substrate fragments of chain length i, mol cm^{-3}
C_s	concentration of substrate in the particle, mol cm^{-3}
C_{sL}	concentration of substrate in the bulk solution, mol cm^{-3}
C_I, C_{II}	concentrations in each phase [Eq. (21)], ML^{-3}
D	diffusion constant, $cm^2\ s^{-1}$
D	inactivated form of enzyme-substrate complex [Eq. (18)], ML^{-3}
D_s	average restricted diffusion coefficient of soluble substrate fragments, $cm^2\ s^{-1}$
D_{SF}	average free diffusion coefficient of the fragments, $m^2\ s^{-1}$
e	enzyme concentration (Eq. (3)], ML^{-3}
E	enzyme
E^*	enzyme-substrate complex concentration, ML^{-3}
E_a	enzyme adsorbed on the surface [Eq. (22)], mol
E_A^A	enzyme fraction active on amorphous cellulose, ML^{-3}
E_A^C	enzyme fraction active on crystalline cellulose, ML^{-3}
E_A^d	deactivated form of enzyme
E_{ads}	adsorbed protein, (mg protein) (mg cellulose^{-1})
$E_{ads,m}$	maximum adsorbed protein, (mg protein) (mg cellulose^{-1})
E_I	number of moles of enzyme on the surface, mol
E_0	protein concentration in the supernatant, mg ml^{-1}
(E_0)	initial enzyme concentration, ML^{-3}
E_1	enzyme concentration (exo-glucanase) [Eq. (10)], ML^{-3}
E^*G	enzyme-substrate-product complex, ML^{-3}
ES	enzyme-substrate complex
EP	enzyme-product complex
f	fraction of amorphous cellulose in the total cellulose, dimensionless
G	glucose concentration, ML^{-3}
G_2	cellobiose concentration
$[G_2]$	cellobiose in polymerized form [Eq. (10)]
$[G_2]_0$	initial cellobiose concentration in polymerized form, mol^{-3}
G^1, G^2, G^3, G^4	glucose, cellobiose, cellotriose, cellotetraose concentration, ML^{-3}
G_x	reducing sugar concentration, ML^{-3}
GG	cellobiose concentration, ML^{-3}
$[I]$	inhibitor concentration, ML^{-3}
k	rate constant, T^{-1}

k', k''	rate constants, T^{-1}
K	Michaelis constant [Eqs. (47) ~ (51)]
K	partition coefficient [Eq. (21)]
k_i	rate constant associated with cellulose component i, T^{-1}
K_i	dissociation constant for the EP complex [Eq. (13)], ML^{-3}
K_i'	modified equilibrium constant between enzyme and products [Eq. (9)]
K_m	Michaelis constant, ML^{-3}
K_m'	modified Michaelis constant, $g\,l^{-1}$
K_p	constant [Eq. (1)], $ml\,mg^{-1}$
K_p	equilibrium constant between enzyme and products
K_s	dissociation constant for the ES complex [Eq. (13)]
$\bar{k}$	overall mass transfer coefficient of the substrate, $cm\,s^{-1}$
$k_1, k_{-1}, k_2, k_3, k_4, k_5, k_{-5}$	rate constants, T^{-1}
K_1	constant, dimensionless
K_3	constant, dimensionless
K_5	constant, dimensionless
k^2, k^3, k^4	rate constants for cellobiose, cellotriose and cellotetraose
$k_{2:2}, k_{3:1}$	rate constants
m	constant, dimensionless
M	mass taken up at the boundary [Eq. (51)], ML^{-2}
n	constant, dimensionless
n	number of particles [Eq. (60)]
P	product (cellobiose) concentration [Eq. (13)], mol
$P, (P)$	product concentration, ML^{-3}
r	radial position within particles, cm
R	particle radius, cm
S	cellulose concentration, ML^{-3}
$(S), [S]$	cellulose concentration, ML^{-3}
$[S]_a$	effective substrate concentration [Eq. (7)], $g\,l^{-1}$
S_a	amorphous cellulose concentration, ML^{-3}
S_c	crystalline cellulose concentration, ML^{-3}
$[S]_i$	substrate concentration at time i [Eqs. (8) and (9)], $g\,l^{-1}$
$[S]_t$	total substrate concentration [Eq. (7)], $g\,l^{-1}$
$[S]_{i+t_r}$	substrate concentration at time $i + t_r$ [Eq. (8)], $g\,l^{-1}$
$S_0, (S)_0, (S_0), [S]_0$	initial cellulose concentration, $g\,l^{-1}$ or $mol\,l^{-1}$
t	time, T
t_r	time required for reducing the cellulose concentration from $[S]_i$ to $[S]_{i+t_r}$, T
v	rate of reaction, $ML^{-3}\,T^{-1}$
V	maximum rate of reaction, $ML^{-3}\,T^{-1}$
V'	modified maximum rate of reaction [Eqs. (8) and (9)], $g\,l^{-1}\,h^{-1}$
v_i	initial reaction rate, $ML^{-3}\,T^{-1}$
V_L	volume of the liquid phase, cm^3
V_I	volume of phase I at equilibrium, L^3
x	distance normal to the surface of reaction, L

X	extent of hydrolysis, %
X_A, X_c	enzyme-substrate complex
X_1, X_2, X_3	enzyme-substrate-product complex
X_{1m}	maximum value of X_1 [Eq. (25)]
X_2	enzyme-substrate complex [Eq. (25)]
X_1	enzyme-substrate complex [Eq. (25)]
X_1	enzyme-crystalline cellulose complex [Eq. (31)]
X_2	enzyme-amorphous cellulose complex [Eq. (32)]
X_3	enzyme-product complex [Eq. (33)]
Y_A, Y_c	enzyme-substrate-product complexes
α	constant, dimensionless
γ	pore radius of particle, cm
η	correction factor pertaining to diffusional restriction, dimensionless
ϕ	constant [Eq. (57)], dimensionless

9 References

1. Amemura, A., Terui, G.: J. Ferment. Technol. *43*, 275 (1965a)
2. Amemura, A., Terui, G.: J. Ferment. Technol. *43*, 281 (1965b)
3. Arminger, W. G., Zabriskie, D. W., Humphrey, A. E., Lee, S. E., Moreira, A. R., Joly, G.: AIChE Symp. Ser. 158, *72*, 77 (1976)
4. Brandt, D., Hontz, L., Mandels, M.: AIChE symp. Ser. *69*, No. 133, 127 (1973)
5. Brown, D. E., Waliuzzaman, M.: In: Proc. Bioconversion Symp. T. K. Ghose (Ed.) p. 351. New Delhi: IIT 1977
6. Cowling, E. B.: In: Biotech. Bioeng. Symp. No. 5, C. R. Wilke (Ed.), p. 163. New York: Interscience 1975
7. Cowling, E. B., Kirk, T. K.: In: Biotech. Bioeng. Symp. No. 6, E. L. Gaden, Jr., M. H. Mandels, E. T. Reese, L. A. Spano (Eds.), p. 95, New York. Interscience 1976
8. Dwivedi, C. P., Ghose, T. K.: J. Ferment. Technol. *57*, 15 (1979)
9. Eriksson, K. E., Hollmark, B. H.: Arch. Biochem. Biophys. *133*, 233 (1969)
10. Fan, L. T., Lee, Y.-H., Beardmore, D. H.: In: Adv. in Biochem. Eng. Vol. 14, A. Fiechter (Ed.), p. 101. Berlin: Springer 1980
11. Fan, L. T., Lee, Y.-H., Beardmore, D. H.: Biotech. Bioeng. *22*, 177 (1980)
12. Ghose, T. K.: Biotech. Bioeng. *11*, 239 (1969)
13. Ghose, T. K., Bisaria, V. S.: Biotech. Bioeng. *21*, 131 (1979)
14. Ghose, T. K., Bisaria, V. S., Dwivedi, C. P.: In: Proced. V. Int. Ferment. Symp., H. Dellweg (Ed.), p. 439, Berlin 1976
15. Ghose, T. K., Das, K.: In: Adv. in Biochem. Eng. Vol. 1., T. K. Ghose, A. Fiechter (Eds.), p. 55. Berlin: Springer 1971
16. Gong, C. S., Ladish, M. R., Tsao, G. T.: Biotech. Bioeng. *19*, 959 (1977)
17. Halliwell, G.: Biochem. J. *79*, 185 (1961)
18. Halliwell, G.: Biochem. J. *95*, 270 (1965)
19. Howell, J. A., Mangat, M.: Biotech. Bioeng. *20*, 847 (1978)
20. Howell, J. A., Stuck, J. D.: Biotech. Bioeng. *17*, 873 (1975)
21. Huang, A. A.: Biotech. Bioeng. *17*, 1421 (1975a)
22. Huang, A. A.: In: Biotech. Bioeng. Symp. No. 5, C. R. Wilke (Ed.), p. 245. New York: Interscience 1975b
23. Huang, A. A.: Unpublished Report of U.S. Army Natick Laboratory, 1975c
24. Karrer, P., Schubert, P.: Helv. Chim. Acta, *9*, 893 (1926)
25. Karrer, P., Schubert, P., Wehrli, W.: Helv. Chim. Acta, *8*, 797 (1925)

26. Kim, C.: In: ARO Report 74-2, Proceedings of the 1974 Army Numerical Analysis Conference, p. 507. The Office of the Chief of Research, Development and Acquisition 1974
27. King, K. W.: Biochem. Biophys. Res. Commun. *24*, 295 (1966)
28. King, K. W.: Arch. Biochem. Biophys. *120*, 462 (1967)
29. Ladish, M. R.: Ph. D. Dissertation, Enzymatic Hydrolysis of Cellulose: Kinetics and Mechanism of Selected Purified Cellulase Component. Purdue University 1977
30. Lee, S. E.: Ph. D. Dissertation, Modeling and Kinetic Studies of Cellulose Biodegradation Reactions. Philadelphia: University of Pennsylvania 1977
31. Lee, S. E., Armiger, W. B., Watteeuw, C. M., Humphrey, A. E.: Biotech. Bioeng. *20*, 141 (1978)
32. Lee, Y.-H., Fan, L. T.: In: Adv. in Biochem. Eng. Vol. 17, A. Fiechter (Ed.), p. 131. Berlin: Springer 1980
33. Li, L. H., Flora, R. M., King, K. W.: Arch. Biochem. Biophys. *111*, 439 (1965)
34. Mandels, M., Kostick, J., Parizek, R.: J. Polymer Sci.: Part C, No. 36, 445 (1971)
35. Mandels, M., Reese, E. T.: In: Advances in Enzymic Hydrolysis of Cellulose and Related Material, E. T. Reese (Ed.), p. 115. New York: Pergamon 1963
36. Mangat, M. S.: Ph. D. Dissertation, A Kinetic Study of the Enzymatic Hydrolysis of Cellulose. State University of New York at Buffalo, Buffalo 1955
37. Marsh, C. A.: Biochim. Biophys. Acta, *122*, 367 (1966)
38. Merchant, M. V.: TAPPI, *40*, 771 (1957)
39. McLaren, A. D.: Enzymologia, *26*, 237 (1963)
40. McLaren, A. D., Packer, L.: In: Advances in Enzymology, F. F. Nord (Ed.), Vol. 33, p. 245. New York: Interscience 1970
41. Miyamoto, S., Nisizawa, K.: J. Veterinary Soc. Army Japan, *396*, 778 (1942)
42. Nisizawa, K., Hashimoto, Y., Shibata, Y.: In: Advances in Enzymic Hydrolysis of Cellulose and Related Material. E. T. Reese (Ed.), p. 171. New York: Pergamon 1963
43. Okazaki, M., Moo-Young, M.: Biotech. Bioeng. *20*, 637 (1978)
44. Peitersen, N., Medeiros, J., Mandels, M.: Biotech. Bioeng. *19*, 1091 (1977)
45. Rautela, G. S., King, K. W.: Arch. Biochem. Biophys. *123*, 589 (1968)
46. Reese, E. T., Mandels, M.: In: Cellulose and Cellulose Derivatives, N. M. Bikales, L. Segal (Eds.), "Series on High Polymer", Vol. 5, Part 5, p. 1079. New York: John Wiley 1971
47. Renkin, E. M.: J. Gen. Physiol. *38*, 225 (1954)
48. Ross, L. W., Updegraff, D. M.: Biotech. Bioeng. *13*, 99 (1971)
49. Simha, R.: J. Appl. Phys. *12*, 570 (1941)
50. Stone, J. A., Scallan, A. M., Donefer, E., Ahlgren, E.: In: Adv. Chem. Series, G. J. Hajny, E. T. Reese (Eds.), *95*, p. 219, American Chem. Soc., Washington D.C. 1969
51. Stuck, J. D.: Ph. D. Dissertation, Enzymatic Hydrolysis of Pure and Waste Cellulose, State University of New York at Buffalo, Buffalo 1973
52. Suga, K., van Dedem, G., Moo-Young, M.: Biotech. Bioeng. *17*, 185 (1975)
53. Suga, K., van Dedem, G., Moo-Young, M.: Biotech. Bioeng. *17*, 433 (1975)
54. Toda, S., Suzuki, H., Nisizawa, K.: J. Ferment. Technol. *46*, 711 (1968)
55. Updegraff, D. M.: Biotech. Bioeng. *13*, 77 (1971)
56. Van Dyke, Jr., B. H.: Ph. D. Dissertation, Enzymatic Hydrolysis of Cellulose. A Kinetic Study, M.I.T., Cambridge 1972
57. Walseth, C. S.: TAPPI, *35*, 228 (1952)
58. Wheatley, M. A., Moo-Young, M.: Biotech. Bioeng. *19*, 219 (1977)
59. Whitaker, D. R.: Arch. Biochem. Biophys. *43*, 253 (1953)
60. Whitaker, D. R.: Arch. Biochem. Biophys. *53*, 439 (1954)
61. Whitaker, D. R.: Canad. J. Biochem. Physiol. *43*, 102 (1956)
62. Whitaker, D. R.: In: Marine Boring and Fouling Organisms, D. L. Ray (Ed.), p. 301. Seattle: University of Washington Press 1959
63. Wilke, C. R., Mitra, G.: In: Biotech. Bioeng. Symp. No. 5, C. R. Wilke (Ed.), p. 253. New York: Interscience 1975
64. Wilke, C. R., Yang, R. D.: In: Symposium on Enzymatic Hydrolysis of Cellulose, M. Bailey, T. M. Enari, M. Linko (Eds.), p. 485. Aulanko, Finland 1975

Bei Fragen zur Produktsicherheit wenden Sie sich bitte an:
If you have any questions regarding product safety,
please contact:

Walter de Gruyter GmbH
Genthiner Straße 13
10785 Berlin
productsafety@degruyterbrill.com